Genetic Mapping in Experimental Populations

Genetic linkage maps are an increasingly important tool in both fundamental and applied research, enabling the study and deployment of genes that determine important biological traits. This concise introduction to genetic mapping in species with disomic inheritance enables life science graduate students and researchers to use mapping software to produce more reliable results. After a brief refresher on meiosis and genetic recombination, the steps in the map construction procedure are described, with explanations of the computations involved. The emphasis throughout is on the practical application of the methods described; detailed mathematical formulae are avoided and exercises are included to help readers consolidate their understanding. A chapter on recognizing and solving problems provides valuable guidance for dealing with real-life situations. An extensive chapter dedicated to the more complex situation of outbreeding species offers a unique insight into the approach required for many economically important and model species, both plants and animals.

Dr J.W. (Johan) Van Ooijen has worked in the field of genetic linkage analysis for more than 20 years. He developed the QTL analysis software MapQTL® and was also involved in the development of JoinMap®, Professor Piet Stam's genetic linkage mapping software. He continues the development of both software packages in his company, Kyazma B.V., and regularly teaches courses on genetic mapping and QTL analysis.

Dr J. (Hans) Jansen is a senior member of the Biometris group at Wageningen University & Research Centre. He is involved in research projects on various applications of molecular markers, including genetic diversity studies, the construction of genetic linkage maps and QTL analysis involving multiple pedigree-related populations.

Genetic Mapping in Experimental Populations

J.W. Van Ooijen
Kyazma B.V., Wageningen, The Netherlands

J. Jansen
Biometris, Wageningen University &
Research Centre, Wageningen, The Netherlands

CAMBRIDGE
UNIVERSITY PRESS

CAMBRIDGE
UNIVERSITY PRESS

University Printing House, Cambridge CB2 8BS, United Kingdom

One Liberty Plaza, 20th Floor, New York, NY 10006, USA

477 Williamstown Road, Port Melbourne, VIC 3207, Australia

314-321, 3rd Floor, Plot 3, Splendor Forum, Jasola District Centre, New Delhi - 110025, India

79 Anson Road, #06-04/06, Singapore 079906

Cambridge University Press is part of the University of Cambridge.

It furthers the University's mission by disseminating knowledge in the pursuit of education, learning and research at the highest international levels of excellence.

www.cambridge.org
Information on this title: www.cambridge.org/9781107601031

First published 2013

A catalogue record for this publication is available from the British Library

Library of Congress Cataloging in Publication data
Ooijen, J.W. Van, 1957–
Genetic mapping in experimental populations / J.W. Van Ooijen, J. Jansen.
p. ;   cm.
Includes bibliographical references and index.
ISBN 978-1-107-01321-6 (hardback) – ISBN 978-1-107-60103-1 (pbk.)
I. Jansen, J. (Johannes), 1952–   II. Title.
[DNLM: 1. Genetic mapping-methods,   2. genetic linkage,   (3. diploid species)]
572.8´6 – dc23   2013005734

ISBN 978-1-107-01321-6 Hardback
ISBN 978-1-107-60103-1 Paperback

Additional resources for this publication at www.cambridge.org/9781107601031

To the memory of Piet Stam

Contents

Preface   *page* xi

1 Introduction   1
    References   8

2 Meiosis and genetic recombination   9
    2.1   The life cycle and the ploidy level   9
    2.2   Two types of cell division   10
    2.3   Two sexes or mating types   10
    2.4   Meiosis   11
    2.5   Following two loci in meiosis   13
        Interchromosomal recombination   13
        Intrachromosomal recombination   14
    2.6   Recombination frequency   17
        Crossover interference   17
        Recombination frequency and physical distance   18
    2.7   Polyploidy   18
    2.8   Sex chromosomes   19
    References   19

3 Estimation of recombination frequencies   21
    3.1   What do we observe? From separate loci to pairs, from phenotype
        to genotype   21
        Genotype and phenotype notation   23
    3.2   The backcross   23
        Estimation with the method of maximum likelihood   26
        The likelihood ratio and the *LOD*   27
        The variance of the estimator $\hat{r}$ in the backcross   28
        Doubled haploids   28
    3.3   The F_2 29
        The F_2 with dominance   33
        One marker with a dominant allele   33
        Coupling phase   34

		Repulsion phase	35
		Practical consequences of dominance	36
	3.4	Advanced population types	37
		Exercises	39
		References	44

4 Determination of linkage groups — 45

4.1	Why determine linkage groups and how?	45
4.2	Test for independent segregation	46
	Example	48
	Generalization	48
4.3	Test for the recombination frequency being equal to $\frac{1}{2}$	49
	Example	50
	Generalization	50
4.4	Some caution	51
4.5	Grouping algorithm	52
4.6	Determining the linkage groups in practice	53
	Exercises	56
	References	57

5 Estimation of a genetic map — 59

5.1	Recombination frequencies and distances	59
5.2	An example experiment	63
5.3	Map estimation by linear regression	65
	Missing genotype observations	68
	Crossover interference	68
5.4	Map estimation by maximum likelihood	69
	Missing genotype observations	71
	Crossover interference	72
	Exercises	72
	References	75

6 Criteria for the evaluation of maps — 77

6.1	Introduction	77
6.2	Log-likelihood	78
6.3	Sum of *LOD*s of adjacent segments	79
6.4	Sum of recombination frequencies in adjacent segments	80
6.5	Residual sum of squares	81
6.6	χ^2 Goodness of fit	82
	Exercises	83

7 How to find the best map order — 87

| 7.1 | Introduction | 87 |

7.2   The travelling salesman problem   88
7.3   Optimization algorithms   90
7.4   Greedy algorithms   91
7.5   Simulated annealing   92
7.6   Tabu search   93
7.7   Evolutionary algorithms   94
7.8   Ant colony optimization   95
7.9   Distance geometry   96
7.10 Final remarks   96
Exercise   96
References   98

8   Outbreeding species   101
8.1   Introduction   101
8.2   Segregation type and linkage phases   102
8.3   Two-way pseudo-testcross strategy   106
        Linkage phase determination in the pseudo-testcross   108
8.4   Estimation of the pairwise recombination frequency in the
        full-sib family   110
        Linkage phase determination in the full-sib family   111
8.5   Simultaneous estimation of the separate parental
        recombination frequencies   112
        Example   114
        Multipoint estimation   116
8.6   Map estimation   117
Exercises   119
References   120

9   Mapping in practice   123
9.1   Introduction   123
9.2   Model assumptions   124
9.3   Observation errors   125
9.4   Phenotype and genotype   127
9.5   Missing observations   128
9.6   Pre-mapping diagnostics   129
        Phenotype frequencies   129
        Segregation ratio   129
        Identical loci and individuals   131
        Pairwise recombination frequencies   131
9.7   Post-mapping diagnostics   132
        Segregation distortion   132
        Number of recombinations per individual   132
        Phenotype probabilities and graphical genotypes   133

Stress	134
Model fit	135
Plausible map orders	136
References	137
Answers to exercises	139
Index	155

Preface

Genetic linkage maps play an important role in many branches of biological research. The specific calculations involved in the construction of genetic maps ask for dedicated computer software. When using this software, the construction of genetic maps looks simple: just press a few buttons and you get a map. However, things are not as simple as they look. The various computations are based on a fair amount of theory from genetics, probability, statistics and optimization. We felt the need to explain this theory and write a special textbook dedicated to the construction of genetic maps. We think it is essential reading for those who want to construct reliable genetic maps.

Ever since the late 1980s, when the first RFLP projects started in Wageningen, we have been involved in linkage analysis with molecular genetic markers. By the mid-1990s, we started teaching short courses on the construction of genetic maps. Currently, our course covers 3 days of lectures and practicals. This book is a complete rewrite of our course reader. The book is aimed at the biologist with an interest in the construction of genetic maps. For a proper understanding, knowledge of genetics, probability and statistics at the undergraduate level is required.

Writing this book would not have been possible without the cooperation of many colleagues, especially the geneticists and plant breeders of Wageningen University and Research Centre. We are very grateful to Dr Hans de Jong, Professor of Cytogenetics at Wageningen University, for his comments and suggestions on the chapter about meiosis.

We dedicate this book to the memory of the late Professor Piet Stam. For a long time, he was a great inspiration to both of us. For J.W.V.O., this started as early as 1976 when he attended Piet's course on population genetics at the Agricultural University in Wageningen. For J.J., it really started in 1990 when Piet joined the group of geneticists and statisticians at the Centre for Plant Breeding and Reproduction Research (now part of Wageningen University & Research Centre). Piet Stam was a very enthusiastic scientist who contributed significantly to genetics. We name just two of his papers: 'Interference in genetic crossing over and chromosome mapping' (*Genetics*, 1979, 92, 573–94), and 'Construction of integrated genetic linkage maps by means of a new computer package: JoinMap' (*The Plant Journal*, 1993, 3, 739–44). We owe much to him for his inspiration and knowledge.

J.W. Van Ooijen & J. Jansen
Wageningen, The Netherlands
7 November 2012

1 Introduction

Around 1900, scientists tried to understand the relationship between the inheritance of simple traits and observations of meiotic cell division under a microscope. It was the time that Mendel's laws on the inheritance of traits (Mendel, 1865) were rediscovered. Mendel did not have any idea of the biological mechanisms underlying his laws. However, some 35 years later, after studying Boveri's 1902 paper, Sutton (1902) realized that chromosomes and their behaviour in meiotic cell division could very well explain Mendel's results. However, in the many experimental crosses carried out to study the simultaneous inheritance of two traits, occasionally large deviations from Mendel's Law of Independent Assortment were observed. Bateson *et al.* (1905) described these deviations in terms of coupling of the heritable factors determining the traits. In subsequent work by Morgan and others, it became clear that heritable factors could be grouped with respect to the law of independent assortment: if two factors belonged to the same group, then their inheritance showed some interdependence; otherwise, the law would hold. In other words, they started realizing that these groups corresponded with chromosomes. By 1911, Morgan showed the possibility of recombination between factors lying on the same chromosome (Morgan, 1911a). Morgan assumed that this was due to an interchange (as he called the crossing over) between homologous chromosomes during meiosis. This corresponded very well with the detailed cytological observations of Janssens (1909). Later, Morgan (1911b) suggested that the heritable factors should be located in a linear fashion on the chromosomes and that the degree of coupling between traits would depend on the distance between the factors on the chromosomes. Sturtevant (1913), a student of Morgan, tested the hypothesis that the proportion of crossovers (as he called the recombinants) could be used as an index of the distance between two factors. He argued that, after determining distances from A to B and from B to C, it should be possible to predict the distance from A to C and to determine the order of the factors on the chromosome. Sturtevant successfully analysed six factors located on the sex chromosome of *Drosophila* and thus became the first ever person producing a genetic map. With his work, the chromosome theory of inheritance became really established.

What exactly is a genetic map? It is a representation of the relative positions of genes and genetic markers on the chromosomes of a biological species. Genes are functional stretches of DNA, whilst genetic markers represent anonymous, non-functional stretches of DNA. Both have fixed positions on the chromosomes, which is why together they are called loci throughout this book. The word 'loci' is the plural of the Latin word 'locus', meaning position. The genetic map of a chromosome (or part of it) is one-dimensional, reflecting the chromosome's linear structure. Distances on a genetic map are measured in units called Morgans, named in honour of the geneticist T. H. Morgan (1866–1945); in practice, centiMorgans are usually used, abbreviated to cM.

Genetic mapping is the procedure for the construction of a genetic map. Genetic linkage analysis is a more general term for any study on the co-segregation of two or more loci. Because genetic maps are always determined with linkage analysis, they are also called linkage maps and even genetic linkage maps. Since Sturtevant's first map, countless genetic maps of many biological species have been produced. In the early years, the goal was to study the chromosomal theory of inheritance, whilst later, relationships in the inheritance of traits became the prime objective. For a long time, mapped traits were simple, morphological characters, whilst later, physiological traits and isozymes were added. One of the major limitations was the limited level of polymorphism encountered with this kind of traits. Without polymorphism, segregation cannot be observed and observations on segregation are a prerequisite for mapping a trait. The development of molecular biology in the 1970s and more specifically the advent of molecular genetic markers started an enormous expansion of genetic linkage mapping. With molecular techniques, we now have an abundance of polymorphic genetic markers. Presently, genetic linkage maps play a prominent role in various fields of fundamental and applied genetic research, for instance, map-based cloning of genes, whole-genome sequencing, quantitative trait loci (QTL) analysis and marker-assisted breeding.

The subject of this book is limited to linkage analysis of experimental populations of species with a regular disomic inheritance. In diploid species, each chromosome has two copies, one inherited from each of the two parents. Diploid species have a disomic inheritance, which is described in Chapter 2 on meiosis. In diploid species, meiosis is the cell division process that produces haploid gametes, which carry a single copy of each chromosome. As mentioned above, segregation is a prerequisite of linkage analysis. If an individual carries different alleles, i.e. variants, at a locus on the two corresponding, homologous chromosomes, this locus will segregate in the offspring of this individual. The essence of genetic linkage analysis is observing how often alleles of loci are inherited together or become exchanged due to crossovers in meiosis. Because the exchange due to a crossover results in new combinations of alleles, this phenomenon is called genetic

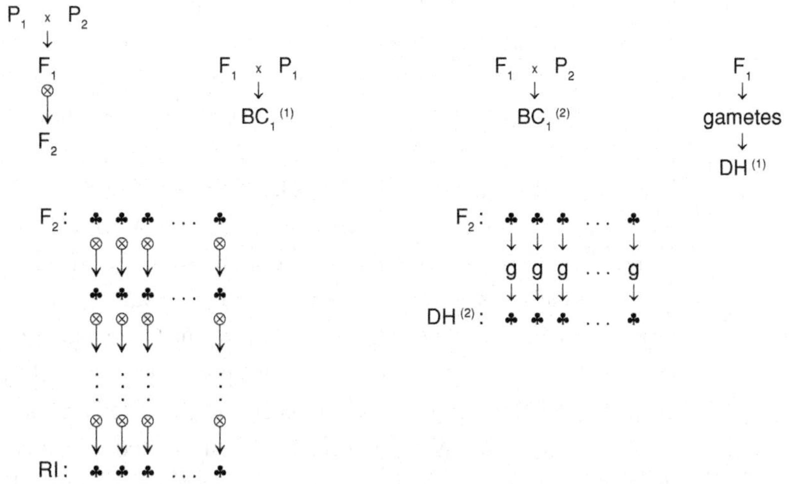

P$_1$ x P$_2$
↓
F$_1$ F$_1$ x P$_1$ F$_1$ x P$_2$ F$_1$
⊗   ↓   ↓   ↓
↓   BC$_1$$^{(1)}$ BC$_1$$^{(2)}$ gametes
F$_2$ ↓
                                                                       DH$^{(1)}$

F$_2$: ♣ ♣ ♣ ... ♣   F$_2$: ♣ ♣ ♣ ... ♣
      ⊗ ⊗ ⊗   ⊗   ↓ ↓ ↓   ↓
      ↓ ↓ ↓   ↓   g g g ... g
      ♣ ♣ ♣ ... ♣   ↓ ↓ ↓   ↓
      ⊗ ⊗ ⊗   ⊗   DH$^{(2)}$: ♣ ♣ ♣ ... ♣
      ↓ ↓ ↓   ↓
      : : :   :
      : : :   :
      ⊗ ⊗ ⊗   ⊗
      ↓ ↓ ↓   ↓
RI: ♣ ♣ ♣ ... ♣

Fig. 1.1

Common experimental population types based on fully homozygous parents, P$_1$ and P$_2$. Crossing the parents results in the heterozygous F$_1$. The F$_1$ can be selfed (⊗) to obtain an F$_2$ population. The F$_1$ can be crossed back to each of the parents, resulting in a first-generation backcross population: BC$_1$$^{(1)}$ or BC$_1$$^{(2)}$. A special treatment of the haploid gametes (g) of the F$_1$ or of each F$_2$ individual results in a doubled haploid population: DH$^{(1)}$ or DH$^{(2)}$. Repeated selfing of each individual by single-seed descent starting from the F$_2$ results in a recombinant inbred line (RI) population. The symbol ♣ represents an individual; the ellipsis (. . .) indicates repetition.

recombination. Two loci are said to be closely linked when there is a strong tendency that the original combinations of alleles are transmitted together to the next generation, i.e. very little genetic recombination occurs. The strength of the linkage is inversely related to the distance between the loci on the chromosomes.

In order to study how often alleles of loci are inherited together or become exchanged by genetic recombination, we have to create segregating progenies (hereafter often referred to as populations). For obvious reasons, this is not an option in human linkage studies; here, segregation must be observed in (large) sets of existing pedigrees. The same applies to linkage studies in certain (livestock) animals. This book is limited to linkage analysis in experimental populations.

An individual carrying different alleles at a certain part of its loci, i.e. it is heterozygous at these loci, will produce different gametes. Therefore, its progeny will consist of different genotypes, in other words, it will reveal segregation for these loci. Figure 1.1 shows a scheme with the most common types of experimental population derived from fully homozygous parent lines, also known as inbred or pure lines. The alleles of a homozygous locus are identical at the homologous chromosomes. A pure line is homozygous over its entire genome. Crossing two distinct pure lines will create an F$_1$ that is heterozygous at all loci at which the two parent lines differ in alleles. The heterozygous F_1 is the starting point for the common segregating population types, such as the first-generation backcross (BC_1) and the F_2. For certain species, it is possible to produce so-called doubled haploids (DH). With a special treatment of the male or female gametes, the haploid set of chromosomes of a gamete is duplicated and subsequently the gamete develops into a diploid, fully homozygous individual. Another common population type is the family of recombinant inbred lines (RI or RIL), which is obtained by repeated inbreeding starting from the F_2 using a process called single-seed descent. Each type of population has its advantages and disadvantages. For instance, in a RIL family and in a DH population, all individuals are homozygous at practically all loci. Because of this, they can be reproduced indefinitely by seeds without changing their genotypes. Such populations are called immortalized populations.

After a segregating population has been created, the next step is to observe the genotypes of the markers, genes and traits of each individual. As already mentioned, genetic linkage maps were originally constructed using mainly morphological markers. Such markers are limited in number; moreover, combinations of alleles at multiple loci may have severe side effects on the resulting phenotypes. Currently, several types of molecular genetic markers are employed. The unique characteristics of the DNA molecule lie at the basis of all molecular marker techniques. Enzymatic cutting of DNA into fragments (also called restriction digestion), electrophoretic separation of fragments, hybridization, ligation and DNA synthesis are the biochemical techniques involved. Molecular markers visualize variation in coding and non-coding parts of the DNA. Molecular variants can be followed genetically, just like the alleles of segregating genes. In molecular size, a molecular genetic marker may represent just a single DNA nucleotide up to a stretch of thousands of nucleotides. As a chromosome consists of billions of nucleotides, molecular markers may be considered as points on chromosomes. Currently, the number of markers employed in a mapping experiment has grown into the thousands for a whole genome. Recent developments in single-nucleotide polymorphism (SNP) technologies are increasing this number dramatically, into the tens of thousands.

With respect to linkage analysis per se, the only difference between the various molecular marker techniques is whether it is possible to identify all possible genotypes of a segregating population. In the case of dominance, the heterozygote cannot be distinguished from one of the homozygotes. Usually, this is caused by the presence of so-called null alleles, which are alleles representing the absence of any (observable) molecular variant. In this book, we do not pay attention to the various molecular marker techniques. However, we want to emphasize that it is important to have at least a basic knowledge of the techniques employed in order to be able to cope with possible problems encountered in linkage analysis.

When all observations on the markers, genes and traits are recorded, the actual linkage analysis begins. As mentioned above, the essence of genetic linkage analysis is observing how often the alleles of loci are inherited together or are exchanged due to genetic recombination. The rate of recombination between two loci is a measure of the distance apart of these loci. Measuring these rates for all pairs of loci is the starting point for constructing the linkage map. In Chapter 3, we deal with the estimation of the recombination frequencies in various common situations. For this purpose, we introduce the method of maximum likelihood, which is a general statistical method for the estimation of parameters.

A genetic linkage map describes the relative positions of genes and genetic markers on the chromosomes. Loci on the same chromosome have a linear arrangement, whilst loci on different (non-homologous) chromosomes are inherited independently. Consequently, no linear relationships exist between the different chromosomes. This is why genetic linkage maps are estimated separately for each chromosome. Determining which loci belong to the same chromosome is therefore a necessary step in the construction of linkage maps. Loci on the same chromosome are physically linked, whereas those on different chromosomes are physically unlinked. Genetic linkage is the phenomenon where traits have a tendency to be inherited together. Due to the independent assortment of the chromosomes in the meiosis, genetic linkage is the result of physical linkage. Determining whether two loci are genetically linked is the starting point of clustering loci into groups. Sets of genetically linked loci are called linkage groups. Chapter 4 tackles the methods for obtaining the linkage groups and describes how to deal with possible complications.

A linkage map is given in linear distance units, whilst the linkage analysis begins with determining recombination frequencies. Therefore, Chapter 5 starts by describing the relationship between recombination frequencies and map distances. Distances can be obtained from recombination frequencies by applying a so-called mapping function. Chapter 5 continues with the description of two methods that can be used to calculate a map for a given order: linear regression and maximum likelihood. The former method is applied to map distances as obtained from the recombination frequencies with a mapping function, whilst the latter is applied to the original genotype observations. The maximum likelihood method estimates recombination frequencies. Because the data of all loci are taken into account, these estimates are often called multipoint estimates of recombination frequencies, in contrast to two-point or pairwise estimates, which are based on pairs of loci only. The multipoint recombination frequencies are subsequently translated into map distances using a mapping function.

Although calculating a genetic linkage map for a given order may be straightforward, the correct order is usually unknown. The order cannot be observed directly from the segregation data, but must be inferred. To do this, we calculate a map for a certain given order and next determine some criterion that expresses the quality of the result. We may repeat this for many possible orders, and identify the best order according to the criterion. In Chapter 6, we present several criteria for the evaluation of maps. Applying these to a small example dataset illustrates that they generally all point to the same order as the best.

As the number of markers increases, the number of possible map orders increases exponentially, so that quite soon it becomes infeasible to evaluate all possible orders. Efficient optimization techniques are required to obtain the best map for a given set of data, thereby circumventing the need to evaluate all possible orders. The techniques must not only be efficient in the sense of computational speed. Fast but poor methods may result in orders that are optimal for parts of the linkage group only. Therefore, the techniques must also be successful in finding orders that are close to the optimal order for the entire linkage group. Chapter 7 is dedicated to optimization algorithms that have been applied to the locus-ordering problem.

The genetic circumstances found in outbreeding species are more complicated than in inbreeding species: (i) more than just two alleles per locus may be present, and (ii) the linkage phases may vary across the loci and between the parents of an experimental cross. Consequently, linkage analysis of outbreeding species actually concerns the analysis of the segregation in two distinct meioses. The two meioses can be analysed separately, which is called the two-way pseudo-testcross strategy. However, it can also be done simultaneously. Both approaches will result in two separate maps, which may be combined into one as an average for both meioses. Chapter 8 explains in detail the genetic situations encountered in outbreeding species and describes the linkage analysis of a full-sib family of an outbreeding species.

We dedicate the final chapter to practical aspects of genetic linkage mapping. The analysis methods presented in this book are based on current knowledge of meiosis. This knowledge is converted into a mathematical model. Experimental data are analysed according to this model resulting in a map. Essentially, a map is used for prediction purposes: in map-based cloning, for indirect selection in breeding or in QTL analysis. Current experience shows that linkage mapping is a very reliable tool if the data to be analysed are of high quality. In practice, however, experimental data may suffer from missing observations and errors. Segregation may occur in a non-Mendelian way, i.e. not according to our model of meiosis. Consequently, the data may not fit the model (or, vice versa, the model is not in accordance with the data). Therefore, prior to the linkage mapping, the data should be inspected carefully in order to assess the quality. Errors may be random or systematic; if they are systematic, any poor-quality marker or individual should be identified and removed. If the poor-quality observations concern the gene of

Table 1.1. Genetic key figures (roughly rounded) for some selected species. *N*, haploid number of chromosomes; TML, total map length (cM); CML, average chromosome map length (cM); bp, number of DNA base pairs ($\times 10^9$); bp/TML, average number of base pairs per cM ($\times 10^6$).

Species	*N*	TML	CML	bp	bp/TML
Arabidopsis	5	600	120	0.15	0.25
Barley	7	1000	140	5	5
Caenorhabditis elegans	6[a]	500	85	0.1	0.2
Lily	12	1000	85	40	40
Maize	10	1700	170	3	1.8
Man	23[a]	3500	150	3	0.9
Onion	8	700	90	10	14
Table mushroom	13	1200	90	0.03	0.03
Tomato	12	1400	120	0.9	0.65

[a] Including the sex chromosome.

interest, removal is not an option, so the quality of the observations will need to be improved. Sometimes, systematic errors may not surface until after a linkage map has been calculated. If so, poor-quality markers or individuals should be identified and removed and the map recalculated. Thus, the mapping procedure can be characterized as an iterative process of pre-mapping diagnostics, actual mapping and post-mapping diagnostics.

What is the relationship between a genetic linkage map and a physical map? A genetic linkage map concerns genetic recombination probabilities, whilst a physical map is about base pairs (or nucleotides) and sequences. Table 1.1 lists some genetic key figures for a few species. You can see that there is quite some variation in the haploid number of chromosomes, but that the average length of the genetic linkage map of a chromosome has a limited range, with an overall average of roughly 100 cM. It usually varies between 50 and 200 cM among and within species. At the same time, a considerable amount of variation is present in the number of base pairs across species. Moreover, a more than 1000-fold range in the number of base pairs is present in a centiMorgan map length. In other words, across species there is no constant one-to-one relationship between the linkage and the physical map. Furthermore, no such constant relationship exists within species: because recombination itself is a biological process, you may expect to find variation among individuals within the same species. Finally, there are so-called hot and cold spots of recombination: on certain physical locations on the chromosomes, relatively many or only a few recombination events take place. This means that, even within an individual, a large amount of variation exists in the numbers of base pairs for each centiMorgan map length.

Linkage mapping requires a basic understanding of genetics, molecular biology, probability, statistics and optimization techniques. The goal of this textbook is to provide the reader with an adequate level of knowledge on these subjects without going into too much mathematical detail. This should enable him or her to produce linkage maps of experimental populations (with the help of computer software). In addition, he or she should be able to recognize and subsequently deal with various problematic situations that may be encountered in practice.

References

Bateson, W., Saunders, E. E. & Punnett, E. C. (1905). Experimental studies in the physiology of heredity. *Reports to the Evolution Committee of the Royal Society. Report II*. http://www.archive.org.

Boveri, T. (1902). Über mehrpolige Mitosen als Mittel zur Analyse des Zellkerns. *Verhandlungen der physicalisch-medizinischen Gesselschaft zu Würzburg. Neue Folge*, 35, 67–90.

Janssens, F. A. (1909). La théorie de la chiasmatypie. Nouvelle interprétation des cinèses de maturation. *La Cellule*, 25, 389–411. http://archive.org/stream/lacellule25lier. Translated to English: *Genetics*, 191 (2012), 319–346.

Mendel, G. J. (1865). *Versuche über Pflanzenhybriden*. Verh. des Naturf. Vereins, Brünn. IV. Band. Abhandlungen 1865, Brünn, 1866. S. 3–47. http://www.biologie.uni-hamburg.de/b-online/e08/08a.htm and http://www.mendelweb.org.

Morgan, T. H. (1911a). The application of the conception of pure lines to sex-limited inheritance and to sexual dimorphism. *American Naturalist*, 45, 65–78. http://www.jstor.org/stable/2455465.

Morgan, T. H. (1911b). Random segregation versus coupling in Mendelian inheritance. *Science*, 35, 384. http://www.jstor.org/stable/1638198.

Sturtevant, A. H. (1913). The linear arrangement of six sex-linked factors in *Drosophila*, as shown by their mode of association. *Journal of Experimental Zoology*, 14, 43–59. http://www.esp.org/timeline.

Sutton, W. S. (1902). On the morphology of the chromosome group in *Brachystola magna*. *Biological Bulletin*, 4, 24–39. http://www.esp.org/timeline.

2 Meiosis and genetic recombination

A genetic linkage map is a representation of the relative positions of genetic loci on the chromosomes. Its construction is based on knowledge of how often alleles of different loci are inherited together or become exchanged by genetic recombination. Therefore, it is important to understand how genetic recombination takes place during meiosis. In eukaryotic organisms, meiosis is a series of two special cell divisions, enabling the creation of new combinations of alleles. The alternative type of cell division, called mitosis, merely produces genetically identical copies of cells. This chapter contains a concise description of meiosis as far as is relevant for the understanding of genetic recombination. Recombination frequency is introduced as a measure of the rate of genetic recombination.

2.1 The life cycle and the ploidy level

A cell containing one basic set of chromosomes is called *haploid*. A cell containing two sets is called *diploid*. In some species, cells contain more than two sets of chromosomes; such species are called *polyploid* and will be discussed briefly in Section 2.7. The life cycle of eukaryotic species is characterized by successions of haploid and diploid phases. In both phases, *mitosis* enables vegetative growth leading to multicellular organisms or it enables multiplication leading to more, genetically identical individuals, as in microorganisms. *Meiosis* is responsible for the transition from the diploid to the haploid phase. At some stage, two haploid cells will merge to form the so-called *zygote*, thereby closing the life cycle at the diploid level. The lengths of the haploid and diploid phases of the life cycle vary greatly among and within higher and lower eukaryotes. For example, for most fungi, the zygote forms the entire diploid phase in the life cycle: the zygote goes directly into meiosis and the resulting haploid cells divide mitotically into multicellular haploid mycelia. In most animals, the four cells produced by meiosis develop into gametes without further cell divisions, and the gametes are ready to fuse with gametes of the opposite sex to form diploid zygotes. Here, the gametes represent the entire, but very short, haploid phase in the life cycle. In most higher plants, the four meiotic daughter cells undergo a few mitotic divisions before the gamete is produced. The life cycle can also be more complicated: for instance, in honey bees the drones (the males) are haploid, the worker bees (the females) are diploid but sterile, and the queen bee is diploid and fertile.

2.2 Two types of cell division

Eukaryotic organisms are capable of two types of cell division: mitosis and meiosis. The mitotic cell division results in two cells that are genetically identical to the parent cell (and to each other). Mitosis enables vegetative growth or multiplication of the organism in the haploid and diploid phases. In meiosis, a diploid parent cell undergoes two successive cell divisions resulting in four haploid daughter cells. In addition to the duplication of all chromosomes during the process, exchanges are made between the two complete sets of chromosomes in such a way that the four daughter cells each contain one complete set of chromosomes. A meiotic daughter cell may develop into a *gamete*. A gamete is a special type of cell capable of fusing with another gamete (usually of the opposite sex), thereby bringing the life cycle to the diploid level again. In some species, the development of a meiotic daughter cell into a gamete is preceded by a few mitotic cell divisions. In other species, for example in most fungi, the haploid phase is the dominant part of the life cycle.

2.3 Two sexes or mating types

In most eukaryotes, two sexes are present. In some species, such as fungi, no differentiation into sexes exists, or the differentiation is limited to genetically determined *mating types* without morphological distinctions. In other species, such as animals and some plants, male and female individuals exist. In many flowering plants, male and female organs are present on the same individual. The female gametes are called *egg cells* and the male gametes *sperm cells*. In diploid eukaryotes, the gametes are haploid. When the sperm cell fertilizes the egg cell, they form a diploid zygote. A zygote may develop into a diploid individual by means of mitotic cell divisions, which at a certain stage is capable of performing meiosis in order to produce haploid daughter cells, which again completes the life cycle. As mentioned before, in some fungi, the zygote goes directly into meiosis without any mitotic cell divisions.

2.4 Meiosis

Meiotic cell divisions form the transition from the diploid to the haploid stage of the eukaryotic life cycle. Diploid cells carry two sets of chromosomes, one set obtained from each of its two parents. In meiosis, haploid cells are produced, each with a basic set of chromosomes. Within each set, some of the chromosomes may be from the first parent and the others from the second parent. In other words, sets of chromosomes in the daughter cells will consist of different combinations of parental chromosomes. Moreover, there is also a mechanism that enables the exchange of parts of chromosomes. Here, we present a concise description of the meiosis; for more details see the cytogenetics textbooks of Burnham (1962) and Sybenga (1992).

The diploid set of chromosomes can be divided into pairs that contain the same set of genes; these are called *homologous* chromosomes, or *homologues*. Chromosomes that are not homologous are called non-homologues. Each chromosome consists of a double-stranded DNA helix, also called a *chromatid*. Both types of cell division, mitosis and meiosis, start with the duplication of chromosomes into two identical *sister chromatids*. The sister chromatids are joined by the *centromere*. The centromere plays an important role in the distribution of the chromatids into the *daughter cells* later on in the process: it is here where the *spindle fibres* are attached with which the chromatids are pulled to opposite sides. In mitosis, the two sister chromatids of each chromosome are divided between two daughter cells. As a consequence, the daughter cells are genetically identical.

Meiosis is totally different and consists of two consecutive cell divisions. Prior to the first meiotic division, the four chromatids of each homologous chromosome pair line up, a process called *synapsis* (Fig. 2.1c). During synapsis, points of exchange are formed at arbitrary positions between non-sister chromatids. This occurs in such a way that one part of a chromatid becomes connected to the complementary part of one of its homologous non-sister counterparts, whilst the reciprocal happens with the remaining parts of these two non-sister chromatids (Fig. 2.1d). These points of exchange are called *crossovers*. Each homologous pair has one or more crossovers. In each crossover, only two non-sister chromatids are involved, but in one crossover a chromatid may exchange with one non-sister chromatid, whereas in another crossover it may exchange with the other non-sister chromatid. In many species, the crossovers can be made visible under a microscope as X-like structures. These are called *chiasmata* (singular: *chiasma*) after the Greek letter χ. In general, a crossover is a relatively rare event. A very rough estimate of the number of crossovers per chromosome pair is two. The four homologous chromatids are then distributed over four cells by means of two

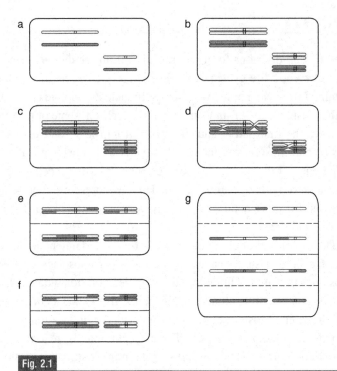

Fig. 2.1

Schematic representation of meiosis. Chromatids are represented by horizontal bars and centromeres by small circles. (a) Prior to meiosis, diploid cells contain two homologues of each chromosome; two chromosomes are shown here. (b) The chromatids are duplicated. (c) The homologous chromatids line up. (d) Points of exchange are formed between non-sister chromatids. (e, f) The chromatids with non-sister centromeres separate and a cell wall is formed between them: this is the first meiotic division. The orientation of the centromeres of one chromosome is independent of the orientation of the other chromosome; therefore, the orientations (e) and (f) are equally likely. (g) In the second meiotic division, the sister centromeres are separated and cell walls are formed between them.

subsequent cell divisions. In the first division, the homologous centromeres are each pulled into the two daughter cells (Fig. 2.1e, f). Next, in the second division, the sister centromeres are each pulled into the final daughter cells (Fig. 2.1g). The distribution of the four chromatids among the four daughter cells takes place at random and independently of non-homologous chromatids. This phenomenon is called the *independent assortment of chromosomes*, and it underlies what is now known as *Mendel's Law of Independent Assortment*.[1]

---

[1]   Mendel's laws. In his original work, *Versuche über Pflanzenhybriden* (1865), Gregor Mendel described the results of his experiments, mainly performed with peas. He described three principles that appeared to determine the heredity of the traits. The first principle says that the F_1 individuals from distinct parents appear uniform, the second says that the F_2 segregates, and the third says that different traits behave independently. In the years after the rediscovery of Mendel's work (~1900), these principles become known as various 'laws'. Unfortunately, the first law became known to some as the Law of Dominance. In his book *Mendel's Principles of Heredity* (1909), William Bateson dismisses this law because

2.5 Following two loci in meiosis

A specific position on a chromosome is referred to as a *locus* (plural: *loci*). At a locus, a functional gene may be present, but it may also be a molecular genetic marker, or both. Variants in the DNA at a locus are called *alleles* of that locus. Loci with different alleles between the homologues are called *heterozygous*; loci with identical alleles are called *homozygous*. In linkage analysis, we are interested in the effects of meiosis on the occurrence of combinations of alleles at different loci. The recombination of alleles due to meiosis can only be observed if loci are heterozygous. Let us consider two loci, **A** and **B**, each with two alleles, (*A*, *a*) and (*B*, *b*), respectively. A common pedigree used in plant genetics starts from two individuals, P_1 and P_2, that are homozygous at all loci: P_1 carries *genotype AA,BB* and parent P_2 carries genotype *aa,bb*. The gametes produced by P_1 are all of type *AB*, whilst those produced by P_2 are all of type *ab*. Obviously, the cross between P_1 and P_2 will lead to a uniform offspring (usually referred to as the F_1), where for all offspring the two loci are heterozygous: *Aa,Bb*. An alternative way of writing genotype *Aa,Bb* is as a combination of two *haplotypes*: *AB/ab*; the word *haplotype* is used to refer to the genetic constitution of a haploid gamete. Now we consider meiosis of the F_1. We need to distinguish between two situations:

1. The loci **A** and **B** reside on different pairs of chromosomes (i.e. on non-homologues).
2. The loci **A** and **B** reside on the same homologous pair of chromosomes.

Interchromosomal recombination

The first situation may result in what is referred to as *interchromosomal* recombination; the process is illustrated in Fig. 2.2. In meiosis, the chromatids are duplicated and distributed over the four daughter cells. For locus **A**, the alleles *A* and *a* are each duplicated into *AA* and *aa* and subsequently distributed over the four haploid daughter cells, leading to two daughter cells with haplotype *A* and two with haplotype *a*. As a consequence, half the daughter cells will obtain the *A* allele, whilst the other half will obtain the *a* allele. This phenomenon is known as *Mendel's Law of Segregation*. The same holds for locus **B**. Because of the independent assortment

it is not always valid: some characters are inherited in an intermediate fashion without dominance. By 1919, Thomas Morgan, in his book *The Physical Basis of Heredity*, ignored Mendel's principle of uniformity as a law and called the principle of segregation 'Mendel's First Law – Segregation of the Genes' and the principle of the independence of traits 'Mendel's Second Law – the Independent Assortment of Genes'. At present, the numbering of the laws varies, some use three and others use just two. It will be less confusing to name rather than number the laws: the Law of Uniformity, the Law of Segregation and the Law of Independent Assortment.

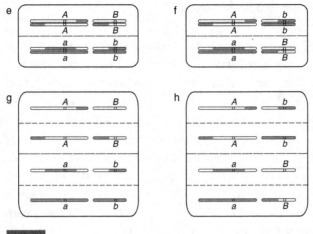

Fig. 2.2

Interchromosomal recombination (continuing from Fig. 2.1). Locus **A** (with alleles *A* and *a*) is located at the centromere of the first chromosome and locus **B** (with alleles *B* and *b*) at the centromere of the second chromosome: they are not affected by crossing over. (e, f) The orientations in (e) and (f) are equally likely due to the independent assortment of chromosomes in the first meiotic division. (g, h) The results after the second meiotic division starting from situations (e) and (f), respectively. Because (g) and (h) are equally likely, all four possible combinations of alleles at loci **A** and **B** are equally likely.

of chromosomes, the distribution of the alleles of locus **A** is independent of that of locus **B**. As a consequence,[2] the meiotic daughter cells have *parental* haplotype *AB*, parental haplotype *ab*, *recombinant* haplotype *Ab* or recombinant haplotype *aB*, each with a probability of 1/4. The parental haplotypes are also referred to as *non-recombinant* haplotypes.

Intrachromosomal recombination

The second situation may result in *intrachromosomal* recombination. Figure 2.3 illustrates what happens in the case of a single crossover. First, the chromatids are duplicated. Then, one crossover occurs between loci **A** and **B** involving exactly two non-sister chromatids, which results in two recombinant haplotypes, *Ab* and *aB*, whilst the two other non-sister chromatids remain unchanged with parental haplotypes *AB* and *ab*. After the distribution of the chromatids among the four daughter cells, two will have parental haplotypes and two will have recombinant

---

[2] If you study only a single meiosis, then you will find one of the following configurations: {*AB AB ab ab*} or {*Ab Ab aB aB*} or {*AB Ab aB ab*}. The latter situation can only occur if at least one of the loci is not located very close to the centromere. The other two situations occur equally frequent, so in a large set of gametes, all four haplotypes are present in equal frequencies.

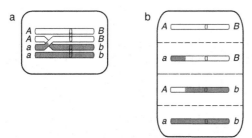

Fig. 2.3

Intrachromosomal recombination when only one crossover occurs between loci **A** and **B**; only one chromosome is shown. (a) Crossover configuration. (b) Distribution of the chromatids among the meiotic daughter cells. All four combinations of alleles at loci **A** and **B** are equally likely.

haplotypes. However, if there is no crossover between **A** and **B**, then two daughter cells will have parental haplotype *AB* and the other two will have parental haplotype *ab*. What happens if there are multiple crossovers between **A** and **B**? In this case, we have to distinguish which chromatids are involved. Suppose there are two crossovers at different positions (Fig. 2.4). Then four possible situations may occur:

1. Two chromatids are involved in the two crossovers. The result will be four non-recombinant daughter cells, because the second crossover cancels the effect of the first crossover.

2 and 3. Three chromatids are involved in the two crossovers. The result will be two recombinant and two non-recombinant daughter cells.

4. All four chromatids are involved in the two crossovers. This results in four recombinant daughter cells.

Normally, there is no preference as to which chromatids are involved in the crossovers[3] and the four situations occur with equal probability. As a consequence, meioses with exactly two crossovers will lead to half the daughter cells having a parental haplotype and the other half having a recombinant haplotype. This result is the same as that for one crossover. It can be shown that with any number of crossovers (except zero), half the daughter cells will have the parental haplotype and half the recombinant haplotype. The conclusion is that the frequency of daughter cells with recombinant haplotypes depends solely on the frequency of meioses in which at least one crossover occurs between loci **A** and **B**.

---

[3] It is assumed that there is no preference as to which chromatids are involved in the crossovers. There are, however, a few rare situations known in which such a preference is detected. This phenomenon is called *chromatid interference*. It may lead to an excess of recombinant haplotypes. See, for example, Stam (1979) and Zhao *et al.* (1995) for a detailed theoretical treatment.

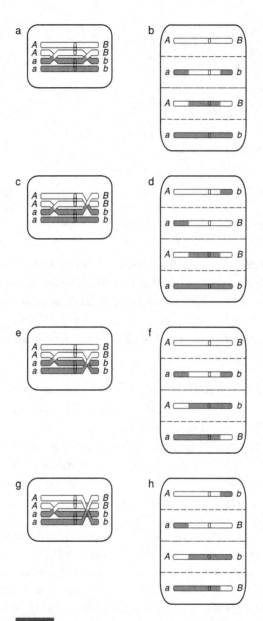

Fig. 2.4

Intrachromosomal recombination when two crossovers occur between two loci **A** and **B**; only one chromosome is shown. Some crossover configurations and the corresponding distribution of the chromatids over the meiotic daughter cells: (a, b) Two chromatids are involved in the crossovers. (c, d) and (e, f) Three chromatids are involved in the crossovers. (g, h) All four chromatids are involved in the crossovers. Overall, all four combinations of alleles at loci **A** and **B** are equally likely.

2.6 Recombination frequency

A simple assumption about the formation of crossovers in the chromosome theory of inheritance is that crossovers occur randomly along the chromosome at a constant rate. This means that if loci lie close together, the probability of a crossover between the loci is small, so that in most meioses no crossover will occur and in most cases all four meiotic daughter cells will have one of the parental or non-recombinant haplotypes. The loci are *linked*. If loci lie far apart, the probability of a crossover is large, and about half of the meiotic daughter cells will have one of the recombinant haplotypes. This leads us to the central idea of linkage analysis: the rate of recombination between two loci is a reflection of the distance between these loci on the chromosomes. Hereafter, the number of recombinant haplotypes relative to the total number of observed haplotypes will be used as the basic measure of linkage. This relative frequency is referred to as the *recombination frequency* (also known as the *recombination fraction*) and usually indicated by the letter r or by the abbreviation RF. The recombination frequency is the primary parameter of interest in linkage analysis.

In the section about interchromosomal recombination, we saw that, for loci on different chromosomes, two out of four meiotic daughter cells have the recombinant haplotype: the corresponding recombination frequency is 0.5. In the section about intrachromosomal recombination, we saw that no recombinant haplotypes occur if there is no crossover between loci, whereas if one or more crossovers occur, then half of the haplotypes will be recombinant. As a consequence, the maximum value of the recombination frequency will be 0.5. The value of 0.5 may already be reached if the loci are far apart on the same chromosome. The minimum recombination frequency is 0, a value that will be attained when loci are very close together.

Crossover interference

In many species, the occurrence of a crossover suppresses the occurrence of another crossover at a nearby position, whereas (very rarely) in some species it is found that the probability of another crossover is increased. These phenomena violate the assumption that crossovers occur independently. They complicate the relationship between recombination counts in neighbouring chromosome segments and the overall joint segment. However, they do not affect the fact that the maximum recombination frequency value is 0.5

Recombination frequency and physical distance

No direct relationship exists between the amount of DNA of a species and the rate at which crossovers occur. For a wide range of species, the number of crossovers is about two per pair of homologues. However, the average amount of DNA per homologue may vary considerably, e.g. from 0.03 billion bp for *Arabidopsis* to 3.3 billion bp for lily. Certain regions on some chromosomes may show more crossovers than others; such regions are called recombination hotspots.

2.7 Polyploidy

Meiosis, as described in this chapter, applies to regular diploid species. However, many species have more than just two basic sets of chromosomes: they are *polyploid*. An important distinction must be made at this point. In some of these polyploid species, the sets of chromosomes are not essentially homologous; the non-homologous but related chromosomes are called *homeologous* chromosomes. In this case, the meiotic behaviour is as described for diploid species, i.e. they have the *disomic* mode of inheritance. Such species are called *allopolyploid*. They are assumed to have evolved from a spontaneous cross between related species, followed by chromosome doubling. Denoting a haploid set of chromosomes with a single letter, this process can schematically be displayed as:

AA × BB → AB → AABB.

Because of their disomic mode of inheritance, linkage analysis in allopolyploids occurs as in diploid species. There may be some difficulties, however, for example if a genetic marker system picks up signals from the homeologous chromosomes. This may complicate determining the genotype at the marker loci involved.

The other type of polyploid is the so-called *autopolyploid*, which has multiple sets of truly homologous chromosomes. Autopolyploids can be seen as having arisen from a multiplication (spontaneous or artificially induced) of complete sets of homologous chromosomes:

AA → AAAA (tetraploid) → AAAAAA (hexaploid).

In the meiosis of autopolyploids, all homologous chromatids may be included in the synapsis. In, for example, the tetraploid potato, this leads to *tetrasomic* inheritance (in general, *polysomic* inheritance), a mode of inheritance far more complex than disomic. In addition, several complicating deviations in chromosome behaviour may occur, such as preferential pairing of chromatids. The linkage analysis of autopolyploids is outside the scope of this book; for further details we refer the reader to the papers by Hackett *et al.* (1998, 2003, 2007), Luo *et al.* (2004), Stift *et al.* (2008) and Leach *et al.* (2010).

2.8 Sex chromosomes

In animal species, a differentiation is usually present between male and female individuals. This phenomenon is typically accompanied by one pair of chromosomes specialized as *sex chromosomes*. In general, the sex chromosomes have evolved into two microscopically distinguishable chromosomes, the larger chromosome called the *X chromosome* and the smaller the *Y chromosome*. In most species, individuals with two X chromosomes (referred to as *homogametic*) are the females, whilst individuals with an X and a Y chromosome (referred to as *heterogametic*) are the males. However, in birds and butterflies it is the other way around. The sex chromosomes are also referred to as the *allosomes* or the *heterochromosomes*, whereas the non-sex chromosomes are called the *autosomes*. Genes residing on the sex chromosomes reveal so-called *sex-linked* segregation: alleles show a strong association with the sex of the individuals. Usually the X and Y chromosome have a homologous segment. In early meiosis of the heterogametic sex, this segment will show synapsis and crossing over can take place. This homologous segment is called the *pseudo-autosomal* region. Butterflies have a very specific deviation. Here, the females (XY) have no crossovers at all, not even on the autosomes. In meiosis of the homogametic sex, crossing over may occur over the entire sex chromosome. In addition to the described X/Y system, there are many variations in the sex chromosome systems; for example, some species have no Y chromosome and the heterogametic sex then has a single X chromosome. Linkage studies in these situations are beyond the scope of this book.[4]

References

Bateson, W. (1909). *Mendel's Principles of Heredity*. London: Cambridge University Press. http://www.archive.org.

Burnham, C. R. (1962). *Discussions in Cytogenetics*. http://www.maizegdb.org/ancillary/Burnham.

---

[4] Because of the corresponding specifics, linkage analysis of sex-linked loci requires some adjustments in the experimental set-up, for instance analysing within one of the sexes only. However, linkage analysis within the differential segment of the Y chromosome is impossible, because there is never a situation where crossovers occur as there are no YY individuals.

Hackett, C. A., Bradshaw, J. E., Meyer, R. C., McNicol, J. W., Milbourne, D. & Waugh, R. (1998). Linkage analysis in tetraploid species: a simulation study. *Genetics Research*, 71, 143–54.

Hackett, C. A., Pande, B. & Bryan, G. J. (2003). Constructing linkage maps in autotetraploid species using simulated annealing. *Theoretical and Applied Genetics*, 106, 1107–15.

Hackett, C. A., Milne, I., Bradshaw, J. E. & Luo, Z. W. (2007). TetraploidMap for Windows: linkage map construction and QTL mapping in autotetraploid species. *Journal of Heredity*, 98, 727–9.

Leach, L. J., Wang, L., Kearsey, M. J. & Luo, Z. (2010). Multilocus tetrasomic linkage analysis using hidden Markov chain model. *Proceedings of the National Academy of Sciences USA*, 107, 4270–4.

Luo, Z. W., Zhang, R. M. & Kearsey, M. J. (2004). Theoretical basis for genetic linkage analysis in autotetraploid species. *Proceedings of the National Academy of Sciences USA*, 101, 7040–5.

Luo, Z. W., Zhang, Z., Leach, L., Zhang, R. M., Bradshaw, J. E. & Kearsey, M. J. (2006). Constructing genetic linkage maps under a tetrasomic model. *Genetics*, 172, 2635–45.

Mendel, G. J. (1865). *Versuche über Pflanzenhybriden*. Verh. des Naturf. Vereins, Brünn. IV. Band. Abhandlungen 1865, Brünn, 1866. S. 3–47. http://www.biologie. uni-hamburg.de/b-online/e08/08a.htm and http://www.mendelweb.org.

Morgan, T. H. (1919). *The Physical Basis of Heredity*. Philadelphia and London: J. B. Lippincott Co. http://www.esp.org.

Stam, P. (1979). Interference in genetic crossing over and chromosome mapping. *Genetics*, 92, 573–94.

Stift, M., Berenos, C., Kuperus, P. & Van Tienderen, P. H. (2008). Segregation models for disomic, tetrasomic and intermediate inheritance in tetraploids: a general procedure applied to *Rorippa* (yellow cress) microsatellite data. *Genetics*, 179, 2113–23.

Sybenga, J. (1992). *Cytogenetics in Plant Breeding*. Berlin: Springer-Verlag.

Zhao, H., McPeek, M. S. & Speed, T. P. (1995). Statistical analysis of chromatid interference. *Genetics*, 139, 1057–65.

3 Estimation of recombination frequencies

The rate of recombination between two loci is a measure of the distance these loci are apart on the chromosomes. Measuring these values for all loci is the starting point for estimating the map of the loci. In Chapter 2, the recombination frequency of two loci was introduced as the measure of genetic recombination: this is the proportion of gametes that are recombinant between the two loci in a single meiosis. A complication with higher plants and animals is that we do not observe the gametes but the diploid individuals in which two gametes are combined. Because in diploids two alleles are present for each locus, there is not always a one-to-one relationship between the observation of a locus and its genotype. Strictly speaking, we should use the term phenotype for observations regarding the genotype. In some cases, it is possible to determine exactly the genotype from the phenotype, for instance in the backcross. In such cases, estimates of recombination frequencies can be obtained by simple counting. We show that the same estimates are obtained by employing the maximum likelihood principle. This same principle can then also be employed in situations where there is no one-to-one relationship between phenotype and genotype. In yet other situations, recombinant genotypes are the result of several subsequent meioses. Here, the observed recombination must be translated into the probability of recombination in a single meiosis.

3.1 What do we observe? From separate loci to pairs, from phenotype to genotype

Recombination is a phenomenon that occurs with respect to pairs of genes or markers. In a regular linkage analysis, however, we start with observing the phenotypes of loci separately. Only as a next step are the observations combined into pairs, which allows the study of recombination. Let us first look at phenotypes: what exactly is a *phenotype*? The term phenotype was introduced by the Danish geneticist W. Johannsen back in 1909 in order to be able to make the distinction between what is observed of an organism and its genetic constitution. By definition, a phenotype is what we can observe of the genotype of an individual. Although somewhat confusing, the pleasant characteristic of many genetic markers is that the phenotype is equal to the genotype. This is why the words are regularly used as synonymous in the context of linkage analysis. However, there can be a major difference.

One quite common case is that of the phenomenon called *dominance*. Diploid individuals possess two alleles of each gene, and if the expression of one of the alleles inhibits the observation of the other allele (i.e. in the phenotype), then the first allele is *dominant* over the second so-called *recessive* allele. The result of this is that it becomes impossible to observe from the phenotype whether the individual has two copies of the dominant allele or a single copy plus a copy of the recessive allele. In other words, in the case of dominance, the heterozygous genotype cannot be distinguished from one of the homozygous genotypes by observing the phenotype. The contrasting situation where the two homozygous and the heterozygous genotypes can be distinguished phenotypically is called *codominance*.

There are situations in which phenotypes are determined by multiple loci on the genome; this phenomenon is called *epistasis* (Bateson, 1909). It is not uncommon for markers in allopolyploid species to have multiple copies on the homeologous chromosomes. Using such markers in linkage analysis will be quite complex. Because the amount of linkage information that can be obtained is very low, employing such markers is generally to be discouraged. Epistasis goes beyond the scope of this book.

Because we wish to observe recombination between pairs of loci, it is important to know which alleles of the loci are coming from the same gamete. For instance, the individual heterozygous at two loci, **A** and **B**, has the phenotype *Aa,Bb*, but there are two genotypes that can explain this phenotype: haplotype combinations *AB/ab* and *Ab/aB*. These genotypes are distinct in the recombination events that generated them. The configuration of these two genotypes is generally described by their so-called *linkage phase*. The linkage phase describes the relationship between alleles at two loci; for the haplotype combination *AB/ab*, the *A* and *B* alleles and the *a* and *b* alleles are said to be in *coupling phase* (sometimes called *cis*-configuration), whereas the *A* and *b* alleles and the *a* and *B* alleles are said to be in *repulsion phase* (sometimes called *trans*-configuration). Of the heterozygous phenotype *Aa,Bb*, the linkage phase is the unknown part with respect to its genotype. However, knowledge of the genotypes of the parents may resolve the linkage phase. For instance, in a testcross or backcross where a heterozygous individual is crossed with a homozygous individual, the homozygous individual obviously produces only one haplotype as gamete and, as a consequence, the linkage phase in any double heterozygous offspring is known.

Genotype and phenotype notation

Before we continue, we wish to make a short side step concerning how genotypes and phenotypes are written, as this may sometimes lead to confusion. In working with diploid species, there are usually two alleles for each locus, especially if just two inbred lines are involved in a cross. For this situation, use is often made of upper- and lower-case characters to represent the two alleles, and different characters for different loci (e.g. alleles A and a for locus **A** and alleles B and b for locus **B**). Sometimes, however, two different characters are used to represent the two alleles of the same locus (e.g. alleles A and B for one locus). When dealing with many loci, you will quickly run out of symbols, and therefore in such situations the same notation is always used for each locus but with a name or number for the locus added somehow. In situations where one allele is dominant over the other, the upper-case character is often used to represent the dominant allele and the lower-case character the recessive allele. This leads to the general misconception that there is dominance involved whenever there are upper- and lower-case characters, but this is not necessarily true, and specifically not in this book. It must be stressed that it is not a convention in the notation in this book that upper-case characters represent dominant alleles and lower-case ones recessive alleles. Therefore, whenever you are dealing with a new situation, you should always make yourself accustomed to the notation used.

In outbreeding species, linkage analysis is often done with crosses of two different heterozygous individuals, where the number of segregating alleles involved can range from two to four, whilst this may vary across the loci. In such cases, it is useful to indicate the type of segregation for each locus and to denote the genotypes with two characters representing the two alleles of an individual. This will be dealt with further in Chapter 8 on outbreeding species.

3.2 The backcross

The simplest pedigree used for linkage analysis is the (first-generation) backcross population for inbreeding species, usually indicated by BC_1. The first linkage map ever produced was made for the sex chromosome in *Drosophila* by Sturtevant (1913) using a scheme in effect similar to that of the backcross. In inbreeding species, genotypes can be propagated by selfing. By selfing for many successive generations, the genotypes become homozygous at all loci. Such fully homozygous genotypes are called *pure lines*. The BC_1 pedigree starts by crossing two pure lines that are homozygous for alternative alleles. We will consider two loci, **A** and **B**, each with two alleles, (A, a) and (B, b), respectively. The lines will be referred to as the first parent, P_1 (with genotype AA,BB), and the second parent, P_2 (with genotype aa,bb). It is important to realize that the two parents will be the *grand*parents of the segregating population, which is two generations down. The cross between the two pure lines produces F_1 individuals that are all identical and heterozygous at loci for which the grandparents have different genotypes, here with haplotype combination AB/ab. In order to obtain the segregating generation, one F_1 individual is *crossed back* with one of the parents; in this case, we use the second parent, P_2. This second generation will be referred to as a backcross or BC_1 population; sometimes the population is referred to as the F_2 *backcross*. Thus, the pedigree of the backcross with regard to loci **A** and **B** is:

Grandparents:	P_1 *AB/AB*	×	P_2 *ab/ab*	
		↓		
Parents:		F_1 *AB/ab*	×	P_2 *ab/ab*
			↓	
Segregating population:			BC_1	

At each locus, the individuals of the BC_1 population are either identical to the backcross parent P_2 (with genotype aa) or to the F_1 (with genotype Aa). The backcross parent P_2 is homozygous and produces only gametes of one haplotype (ab). Therefore, differences between genotypes of individuals of the BC_1 population are solely due to differences between the gametes produced by the F_1 individual. As a result, the backcross enables us to study recombination during gamete formation in a single meiosis (i.e. that of the F_1) and estimate the recombination frequency in a straightforward manner. The F_1 produces recombinant gametes with the probability being the recombination frequency, r. Because there are two equally likely genotypes of recombinant gametes, Ab and aB, each will have a probability of $r/2$. The other gametes are non-recombinant, with genotypes AB and ab, and each will have a probability of $(1-r)/2$. Combining these four possible genotypes of the gametes with the single genotype of the gametes of P_2 gives us the four possible genotypes of the backcross offspring and their probabilities, as shown in (3.1).

Suppose the backcross population consists of N individuals. We will observe the phenotypes of the individuals regarding the two loci. Because of the specific design of the cross, we know that the double heterozygous phenotype Aa,Bb must be of the haplotype combination AB/ab. Thus, for all phenotypes, the genotype is

Genotype	Recombinant status	Probability
AB/ab	Non-recombinant	$(1-r)/2$
ab/ab	Non-recombinant	$(1-r)/2$
Ab/ab	Recombinant	$r/2$
aB/ab	Recombinant	$r/2$
Total		1

(3.1)

completely known. Suppose we observe the following frequencies, n_i, of the four genotypes:

Genotype	Observed frequency
AB/ab	n_1
ab/ab	n_2
Ab/ab	n_3
aB/ab	n_4
Total	N

(3.2)

The total number of recombinant individuals observed is:

$$R = n_3 + n_4.$$

Given that the probability of recombination is r, we would expect to find $E(R) = rN$ recombinant individuals (Remark: $E()$ means the statistical expectation, i.e. the long-term average). If we wish to estimate the recombination frequency, r, we can take that value of r for which the observation R is equal to its expectation:

$$E(R) = R;$$

thus:

$$rN = R.$$

Solving the equation for r gives:

$$\hat{r} = R/N.$$

(3.3)

In words, the estimator of the recombination frequency is the observed relative frequency of recombinant genotypes. Because in situations other than the backcross the estimation is not as straightforward, we now will follow a more formal approach, i.e. that of the method of maximum likelihood. This will lead to exactly the same estimator in the backcross.

Estimation with the method of maximum likelihood

The method of maximum likelihood is a general statistical method for estimating parameters. The approach is based on the *likelihood* (or likelihood function), which is defined as the probability of the data given the probability model that applies to the studied situation. The basic idea behind the method is that the specific value of the parameter to be estimated, for which the likelihood attains its maximum value, will provide a good-quality estimate of the parameter. In order to achieve this, the likelihood function L, which will be a function of the unknown parameter x, $L(x)$, is differentiated with respect to x and equated to zero and the resulting equation is solved for x. Because the maximum of $L(x)$ and the maximum of the natural logarithm of $L(x)$ are determined by the same value of x, and because differentiating this so-called *log-likelihood* $\ell = \ln(L(x))$ is generally much easier than differentiating $L(x)$, usually the log-likelihood is employed.

In our case of the backcross, we first have to define the likelihood based on the applicable probability model. The probabilities of the four possible genotypes are given above (3.1). Of each genotype, we observed n_i individuals (3.2). In our observations, there is no order in the set of genotypes. The number of ways to order the set of observations is equal to the multinomial coefficient:

$$MC = N!/(n_1! \, n_2! \, n_3! \, n_4!).$$

Thus, the likelihood $L(r)$ with respect to the unknown parameter r, given the n_i of observed individuals, is:

$$L(r) = MC \left(\tfrac{1}{2}\right)^{n_1 + n_2 + n_3 + n_4} (1 - r)^{n_1} (1 - r)^{n_2} \, r^{n_3} \, r^{n_4}$$
$$= MC \left(\tfrac{1}{2}\right)^{N} (1 - r)^{n_1 + n_2} \, r^{n_3 + n_4} = MC \left(\tfrac{1}{2}\right)^{N} (1 - r)^{N-R} \, r^{R}.$$

Because MC and $\left(\tfrac{1}{2}\right)^{N}$ are constant values that do not depend on the parameter r, they are usually left out of the likelihood. The log-likelihood, ℓ, is then proportional to:

$$\ell = \ln(L(r)) \propto \ln((1 - r)^{N-R} \, r^{R}) = (N - R)\ln(1 - r) + R \ln(r).$$

The derivative of ℓ with respect to r is:

$$\delta\ell/\delta r = -(N - R)/(1 - r) + R/r.$$

The maximum can be obtained by setting this derivative equal to zero and by solving the resulting equation for r:

$$\delta\ell/\delta r = -(N - R)/(1 - r) + R/r = 0 \Leftrightarrow (N - R)/(1 - r) = R/r;$$

thus, the maximum likelihood estimator of r becomes:

$$\hat{r} = R/N.$$

Here we see that the maximum likelihood estimator is equal to the intuitive estimator that was derived in (3.3). The most important statistical properties of maximum likelihood estimators are consistency and efficiency. Consistency means that when the number of observations is large, then the estimator of a parameter tends to its true value. Efficiency means that when the number of observations is large, then the mean square error of the estimator is minimal.

The likelihood ratio and the LOD

If we wish to test whether two loci are linked or not, we may use the likelihood ratio statistic, LR. If the loci are linked, the recombination frequency is as obtained with the maximum likelihood estimator above. If they are not linked, the recombination frequency is expected to be $\frac{1}{2}$. The likelihood ratio is defined as the ratio of the maximum likelihood as obtained in the previous section and the likelihood obtained under the assumption of 'no linkage', i.e. with $r = \frac{1}{2}$. The formula for the likelihood ratio is:

$$LR = \frac{L(r = \hat{r})}{L\left(r = \frac{1}{2}\right)} = \frac{MC\left(\frac{1}{2}\right)^N (1 - R/N)^{N-R} (R/N)^R}{MC\left(\frac{1}{2}\right)^N \left(1 - \frac{1}{2}\right)^{N-R} \left(\frac{1}{2}\right)^R}$$

$$= 2^N (1 - R/N)^{N-R} (R/N)^R.$$

For testing, the so-called *deviance* is used, which is twice the natural logarithm of the likelihood ratio. Under the null hypothesis (no linkage) and for large values of N, the deviance, D, follows a χ^2 distribution with one degree of freedom:

$$D = 2\ln(LR) \approx \chi_1^2.$$

If linkage is more likely than no linkage, the deviance will obtain a large value. In genetics, often the *logarithm of odds* or LOD is used instead of the deviance, which is defined as the logarithm to the base 10 of the likelihood ratio:

$$LOD = \log_{10}(LR).$$

The interpretation of *the LOD* is as follows. If an estimate of the recombination coefficient is obtained with a *LOD* value equal to 3, then $r = \hat{r}$ is $10^3 = 1000$

times more likely than $r = \frac{1}{2}$. Originally, as a rule of thumb, this *LOD* threshold value of 3 was taken as evidence for linkage. Experience with modern experiments with large numbers of markers shows that this is often not sufficiently stringent, especially when there are many chromosome pairs.

As we will see in the next chapter, the likelihood ratio statistic may be used as a test for establishing the linkage groups that are supposed to represent the chromosomes.

The variance of the estimator $\hat{r}$ in the backcross

According to likelihood theory, the approximate variance of a parameter can be obtained as the inverse of the so-called Fisher information, which is the expectation (E) of minus the second derivative of the log-likelihood. For $\hat{r}$, it follows that:

$$E\left(-\frac{d^2\ell}{dr^2}\right) = E\left(\frac{R}{r^2} + \frac{N-R}{(1-r)^2}\right) = \frac{N}{r} + \frac{N}{1-r} = \frac{N}{r(1-r)},$$

and thus:

$$\mathrm{var}(\hat{r}) = \frac{1}{E\left(-\dfrac{d^2\ell}{dr^2}\right)} = r(1-r)/N.$$

In this case for the backcross, the variance is equal to the variance of a binomial distribution with parameters N and r. An unpleasant property is that, whenever the recombination frequency is estimated as zero, the accompanying estimate of the variance, given by:

$$\hat{\mathrm{var}}(\hat{r}) = \hat{r}(1-\hat{r})/N$$

is also zero. However, if the estimate of r is zero, the true value may still be more than zero, which means that the estimated zero variance gives a false sense of accuracy.

Doubled haploids

For various species, it is possible to give either the male or the female gametes (i.e. cells in the haploid phase of the life cycle) a treatment that results in cells in which each chromosome is duplicated. These cells are then diploid, and because of the identity through the duplication, each cell is completely homozygous. The treatment is followed by growing plants from individual cells. Often this technique is applied to the F_1 of a cross between pure lines. When studying recombination in a family of doubled haploids derived from a single F_1, essentially you are studying recombination in the gametes of this F_1. Because this is also what occurs in studying recombination in the backcross, linkage analysis in such a doubled haploid population is identical to that in the backcross. The main advantage of this type of population is that it will be possible to continue to reproduce identical genotypes of each individual in the population through seeds. Such a type of population is sometimes referred to as an *immortalized* population.

3.3   The F_2

The F_2 is a commonly used pedigree in linkage analysis. It is somewhat more complex than the backcross. The segregating F_2 population is obtained by taking the same F_1 as used in the backcross (i.e. the result of crossing two pure lines) and crossing it with itself. Alternatively, it can be made by crossing one F_1 individual with another F_1 individual of the same cross (remember that all progeny of crossing two pure lines have an identical genotype). The population is sometimes called the *F_2 intercross*. Whereas in the backcross only one parent is heterozygous and creates segregation, in the F_2 both parents are heterozygous and thus the pedigree produces more segregation information. Typical of the pedigree is that both parents of the segregating population have the same genotype (and are often the same individual). Schematically, the pedigree of the F_2 regarding loci **A** and **B** (as described for the backcross) is ($\otimes$ is the symbol for selfing):

Grandparents:	**P₁** *AB/AB*	×	**P₂** *ab/ab*
		↓	
Parents:		**F₁** *AB/ab*	
		⊗	
		↓	
Segregating population:		**F₂**	

At each locus, the individuals of the F_2 population can be homozygous like one of the parents (with genotype *AA* or *aa*) or heterozygous like the F_1 (with genotype *Aa*). To derive the probabilities of all possible two-locus genotypes, it is convenient to start with the so-called *Punnett square* (after Punnett, 1905), which presents the combination of male and female gametes. Because male gametes are combined at random with the female gametes, the probabilities of the diploid genotypes may be calculated as the product of the probabilities of the haploid genotypes of the gametes. For convenience, we define the probability for no recombination as: $s = 1 - r$. At the two loci **A** and **B**, the F_1 generates gametes with four haploid genotypes and acts as both the male and the female parent; the probability (P) of each haploid gamete genotype (G) and combined as diploid F_2 offspring genotype are:

Male		Female							
		G	P	G	P	G	P	G	P
G	P	AB	s/2	ab	s/2	Ab	r/2	aB	r/2
AB	s/2	AB/AB	$s^2/4$	ab/AB	$s^2/4$	Ab/AB	rs/4	aB/AB	rs/4
ab	s/2	AB/ab	$s^2/4$	ab/ab	$s^2/4$	Ab/ab	rs/4	aB/ab	rs/4
Ab	r/2	AB/Ab	rs/4	ab/Ab	rs/4	Ab/Ab	$r^2/4$	aB/Ab	$r^2/4$
aB	r/2	AB/aB	rs/4	ab/aB	rs/4	Ab/aB	$r^2/4$	aB/aB	$r^2/4$

Although the table lists 16 offspring genotypes as combinations of haplotypes, several cells are identical because the order of the haplotypes cannot be determined, e.g. *aB/AB* equals *AB/aB*. Thus, the table is symmetrical across the main diagonal. Furthermore, because the haplotypes cannot be observed but only the genotypes per locus, this means the four cells in which both loci are heterozygous (*AB/ab* and *Ab/aB*) are to be treated as a single observable phenotype *Aa,Bb*. In practice, observations are made on a locus-by-locus basis. For this purpose, we can reorganize the data of the above table to the following table for the unique two-locus genotypes and their corresponding probabilities (*MP*, marginal probability):

(3.4)

Locus A	Locus B						MP
	BB		Bb		bb		
AA	AA,BB	$s^2/4$	AA,Bb	2rs/4	AA,bb	$r^2/4$	$\frac{1}{4}$
Aa	Aa,BB	2rs/4	Aa,Bb	$2(r^2 + s^2)/4$	Aa,bb	2s/4	$\frac{1}{2}$
aa	aa,BB	$r^2/4$	aa,Bb	2rs/4	aa,bb	$s^2/4$	$\frac{1}{4}$
MP	$\frac{1}{4}$		$\frac{1}{2}$		$\frac{1}{4}$		

Note that the marginal probabilities conform to the Mendelian F_2 frequencies of a single locus (*Remark:* $r^2 + 2rs + s^2 = (r + s)^2 = 1$). With regard to the estimation of the recombination frequency in the F_2, two problems arise. First, because the pedigree of the F_2 is symmetrical, it is not possible to distinguish between recombinations in the maternal and paternal meioses. As a consequence, it is only possible to estimate a common (average) recombination frequency for the maternal and paternal meioses. Secondly, if both loci are heterozygous (*Aa,Bb*, a so-called *double heterozygote*), it is not possible to distinguish between situations with no recombinations (*AB/ab*) and with two recombinations (*Ab/aB*). The effect is that it is not possible to estimate the recombination frequency simply by counting recombination events, as can be done in the backcross. The only way out of this problem would be to grow F_3 lines of the double heterozygotes, determine their marker genotypes, and use that information to distinguish whether the F_2 was a double heterozygote in coupling (*AB/ab*, no recombinations) or in repulsion phase (*Ab/aB*, two recombinations). This approach is seldom taken, because the problem is not as dramatic as it may seem.

For the estimation, we would like to apply the method of maximum likelihood. However, a straightforward approach, as with the backcross, unfortunately leads to a polynomial equation without an explicit solution. Nevertheless, a maximum likelihood solution can be obtained quite easily with the so-called *expectation maximization* (EM) algorithm (Dempster *et al.*, 1977). This algorithm is an iterative method to obtain maximum likelihood estimates in situations when there is incomplete information. Each iteration consists of two consecutive steps: in the first step (the E-step), the expectation is determined regarding the incomplete information, whilst in the second step (the M-step), the maximum likelihood estimate is computed under the assumption that the incomplete information is fixed (constant) as expected in the first step. Because of this, the maximum likelihood equation is much easier to solve. With the results of the M-step, a new E-step can be taken, followed by a new M-step, and so on. Under most circumstances, the algorithm will converge towards the maximum likelihood estimates of the unknown parameters.

To set up the likelihood for the F_2 with respect to the recombination frequency, it is more convenient if we consider the two gametes that generated each genotype. Both gametes may contain a recombination regarding the loci **A** and **B**, because the heterozygous F_1 is both male and female parent. The probability of the genotype of an individual can be written as the product of the gamete haplotypes that generated the individual multiplied by a constant ($\frac{1}{2}$ or $\frac{1}{4}$, which will disappear in the differentiation of the log-likelihood) (see 3.4). The probability of a recombinant gamete haplotype is, by definition, r, and that for a non-recombinant haplotype is $s = 1 - r$. Each of the genotypes homozygous at the two loci were created from two non-recombinant (*AA,BB* and *aa,bb*) or two recombinant (*AA,bb* and *aa,BB*) gametes; thus their probabilities are (ignoring the constants) $ss = s^2$ and $rr = r^2$, respectively. The genotypes homozygous at one and heterozygous at the other locus (*AA,Bb*, *aa,Bb*, *Aa,BB* and *Aa,bb*) were created from one non-recombinant and one recombinant gamete each, and thus their probabilities are (ignoring the constants) rs. Finally, for the double heterozygote (*Aa,Bb*), we do not know whether its haplotypes involved zero recombinations or two recombinations, i.e. coupling

(*AB/ab*) or repulsion (*Ab/aB*) phase. This aspect is the incomplete information with respect to the EM algorithm. Suppose we know that the proportions in coupling and repulsion phase are p_{co} and p_{re}, respectively (with $p_{co} + p_{re} = 1$), then we would be able to classify the *average* double heterozygote in the likelihood: for the fraction in coupling phase, the probability is $ss = s^2$, and for the fraction in repulsion phase, the probability is $rr = r^2$. Thus, the likelihood (ignoring the constants) for the entire F_2 is:

$$L(r) = s^S r^R = (1-r)^S r^R,$$

which is similar to the backcross, but here with S being the number of non-recombinant gametes/haplotypes:

$$
\begin{aligned}
S = \quad &2 \quad n_{AA,BB} \quad + \quad 1 \quad n_{AA,Bb} \quad + \quad 0 \quad n_{AA,bb} \quad + \\
&1 \quad n_{Aa,BB} \quad + \quad 2 p_{co} \, n_{Aa,Bb} \quad + \quad 1 \quad n_{Aa,bb} \quad + \\
&0 \quad n_{aa,BB} \quad + \quad 1 \quad n_{aa,Bb} \quad + \quad 2 \quad n_{aa,bb},
\end{aligned}
$$

where n_g are the observed frequencies of each genotype, and with R being the number of recombinant gametes/haplotypes:

$$
\begin{aligned}
R = \quad &0 \quad n_{AA,BB} \quad + \quad 1 \quad n_{AA,Bb} \quad + \quad 2 \quad n_{AA,bb} \quad + \\
&1 \quad n_{Aa,BB} \quad + \quad 2 p_{re} \, n_{Aa,Bb} \quad + \quad 1 \quad n_{Aa,bb} \quad + \\
&2 \quad n_{aa,BB} \quad + \quad 1 \quad n_{aa,Bb} \quad + \quad 0 \quad n_{aa,bb}.
\end{aligned}
$$

The formulae of S and R are presented here with coefficients 0, 1 and 2 at the observed frequencies of each genotype as shown in (3.4) above. These coefficients directly reflect the number of non-recombinant or recombinant haplotypes that underlie each diploid F_2 genotype. Similar to the backcross, the log-likelihood:

$$\ell = ln(L(r)) = S \ln(1-r) + R \ln(r)$$

can be maximized straightforwardly by differentiation if p_{co} and p_{re} are treated as constants, despite the fact that they are functions in r. Thus, the maximum likelihood estimator of r becomes:

$$\hat{r} = R/(S+R) = R/(2N), \tag{3.5}$$

where N is the total number of individuals (Σn_g). The estimator depends on the unknown proportion p_{re} as R depends on it. The proportion p_{re} (i.e. the conditional probability for the repulsion phase given that the genotype is double heterozygote) can be calculated once the recombination frequency is known:

$$p_{re} = r^2/(s^2 + r^2) \text{ [and thus: } p_{co} = s^2/(s^2 + r^2) = 1 - p_{re}\text{].}$$

The iterative solution with the EM algorithm is as follows: the expected value of p_{re} is calculated from the current parameter value r, and then a new value of the parameter r is obtained with the maximum likelihood estimator (3.5) using the current expected value of p_{re}; with the new value of r an improved expectation of p_{re} is calculated, and so on. This should be repeated until there is no more change in the value of r. This is usually reached in fewer than ten iterations. This final value is taken as the maximum likelihood estimate of r. The iterations need an initial value of r to start with; the value $r = \frac{1}{4}$ is usually a successful starting point.

The F$_2$ with dominance

As described at the beginning of this chapter, dominance is a situation where the heterozygous genotype cannot be distinguished from one of the two homozygous genotypes. With molecular markers, this is usually the result of one of the alleles being a so-called *null allele*, i.e. with the technique used, the allele produces no signal; for instance, there is no DNA fragment detected with electrophoresis. If we define '+' as the allele with a signal (the dominant allele) and '−' as the allele without a signal (the recessive allele), then the heterozygous genotype '+ −' cannot be distinguished from the homozygous '+ +', as both produce one and the same signal. In a backcross, this situation either leads to a non-informative marker ('+ −' × '+ +'), or it generates complete information ('+ −' × '− −') regarding the meiosis of the F$_1$, depending on which parent is used to cross back to. In an F$_2$, where both meioses segregate, dominant markers provide less information regarding recombination than codominant markers. For estimating recombination between two markers in an F$_2$ if there is dominance involved, there are three cases to consider. The first case is where only one of the two markers has a dominant allele. Next are two distinct cases where both markers have dominant alleles. In one case, the dominant alleles for both markers are present in one parent and the null alleles in the other. This situation results in the dominant alleles being in coupling phase in the F$_1$. In the other case, the dominant allele for one marker is in one parent, whereas for the other marker it is in the other parent. This results in repulsion phase for the dominant alleles in the F$_1$. We will consider the three cases separately.

One marker with a dominant allele

If in our scheme of the loci **A** and **B** the B allele is a dominant allele, then the Bb genotype cannot be distinguished from the BB genotype. We indicate the corresponding phenotype as $B-$. Thus, the original table of 3×3 two-locus genotypes (3.4) collapses to the following 3×2 table:

Locus A	Locus B					MP
	$B-$			bb		
AA	$AA,B-$	$s^2/4 + 2rs/4 = (1-r^2)/4$		AA,bb	$r^2/4$	$\frac{1}{4}$
Aa	$Aa,B-$	$2rs4 + 2(r^2 + s^2)/4$ $= (2-2rs)/4$		Aa,bb	$2rs/4$	$\frac{1}{2}$
aa	$aa,B-$	$r^2/4 + 2rs/4 = (1-s^2)/4$		aa,bb	$s^2/4$	$\frac{1}{4}$
MP	$\frac{3}{4}$				$\frac{1}{4}$	

For estimation of the recombination frequency, there is no explicit solution to the likelihood equation. Similar to the codominant situation of the F_2 above, the EM algorithm can be employed here. In the codominant F_2, we used expected fractions of the double heterozygote in coupling and repulsion phase in the EM algorithm. Here, there are three phenotypes that have multiple underlying genotypes. Similar to the codominant F_2, for every phenotype representing multiple genotypes, expected fractions must be used to set up the appropriate EM algorithm. Alternatively, other general numerical methods may be applied to obtain estimates of the recombination frequency.

Coupling phase

If in our scheme of the loci **A** and **B** the A and the B alleles are the dominant alleles, then the Aa and Bb genotypes cannot be distinguished from the AA and BB genotypes, respectively. The dominant alleles are in coupling phase in the F_1. We indicate the corresponding phenotypes as $A-$ and $B-$. Thus, the original table of 3×3 two-locus genotypes (3.4) collapses to the following 2×2 table:

Locus A	Locus B					MP
	$B-$			bb		
$A-$	$A-,B-$	$s^2/4 + 2rs/4 +$ $2rs/4 + 2(r^2 + s^2)/4$ $= (2+s^2)/4$		$A-,bb$	$r^2/4 +$ $2rs/4$ $= (1-s^2)/4$	$\frac{3}{4}$
aa	$aa,B-$	$r^2/4 + 2rs/4 = (1-s^2)/4$		aa,bb	$s^2/4$	$\frac{1}{4}$
MP	$\frac{3}{4}$				$\frac{1}{4}$	

For the estimation of the recombination frequency, there is an explicit solution to the likelihood equation (Mather, 1951). Using two auxiliary variables, λ and θ, the explicit maximum likelihood estimator of r becomes:

$$\lambda = (n_{A-,B-} - 2(n_{A-,bb} + n_{aa,B-}) - n_{aa,bb})/(2N)$$

$$\theta = \lambda + \sqrt{(\lambda^2 + 2n_{aa,bb}/N)}$$

$$\hat{r} = 1 - \sqrt{\theta}.$$

Alternatively, an EM algorithm may be set up using a similar approach to that above using expected fractions for underlying genotypes of the phenotypes. Note that phenotype $A-,B-$ has five underlying genotypes.

Repulsion phase

If the A and the b alleles are the dominant alleles of the loci **A** and **B**, respectively, then the Aa and Bb genotypes cannot be distinguished from the AA and bb genotypes, respectively. The dominant alleles are in repulsion phase in the F_1. We indicate the corresponding phenotypes as $A-$ and $b-$. Thus, the original table of 3×3 two-locus genotypes (3.4) collapses to the following 2×2 table:

Locus A	Locus B			MP
	BB		$b-$	
$A-$	$A-,BB$	$s^2/4 + 2rs/4 = (1-r^2)/4$	$A-,b-$ $2rs/4 + r^2/4 + 2(r^2+s^2)/4 + 2rs/4 = (2+r^2)/4$	$\frac{3}{4}$
aa	aa,BB	$r^2/4$	$aa,b-$ $2rs/4 + s^2/4 = (1-r^2)/4$	$\frac{1}{4}$
MP		$\frac{1}{4}$	$\frac{3}{4}$	

Similar to the coupling phase situation, there is an explicit and an EM maximum likelihood estimator for the recombination frequency (Mather, 1951). Using the two auxiliary variables λ and θ, the explicit maximum likelihood estimator of r is:

$$\lambda = (n_{A-,b-} - 2(n_{A-,BB} + n_{aa,b-}) - n_{aa,BB})/(2N)$$

$$\theta = \lambda + \sqrt{(\lambda^2 + 2n_{aa,BB}/N)}$$

$$\hat{r} = \sqrt{\theta}.$$

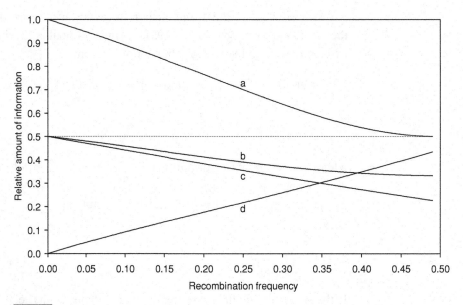

Fig. 3.1

Relationship between recombination frequency and the information relative to the completely classified F_2 for the F_2 with (a) two codominant loci, (b) a codominant locus with a dominant locus, (c) two dominant loci in coupling phase and (d) two dominant loci in repulsion phase.

Practical consequences of dominance

Although it is possible to provide estimates of the recombination frequency under the three situations with dominance involved, it is interesting to consider the accuracy. It will be obvious that information regarding the segregation is lost due to dominance: not all genotypes can be observed directly, and Mendelian assumptions have to be made regarding the distributions of genotypes underlying the observed counts of phenotypes. Figure 3.1 (after Mather, 1936) shows the Fisher information of these situations relative to the fully informative situation. In the fully informative situation, there are two codominant loci and additionally the distinction is made between coupling and repulsion phases of the double heterozygote, so that all ten genotypes of the F_2 can be observed directly. From this figure, it is clear that dominance leads in all three situations to a significant reduction in the amount of information. In particular, the case of two dominant loci in repulsion phase provides a very low amount of information. One of the manifestations of this characteristic is that very often the recombination frequency will become estimated as zero, whereas the real value may be up to 0.2 or even larger. This can be derived from considering the situation where the only class with direct information about recombination, *aa,BB*, has zero observations. It can

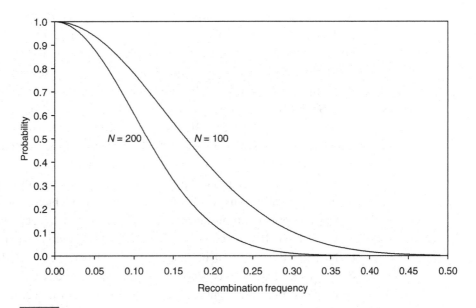

Fig. 3.2

Probability of obtaining a count of zero individuals of genotype *aa,BB* in the F$_2$ with two dominant loci in repulsion phase, at population sizes of $N = 100$ and $N = 200$.

be shown that, when this *aa,BB* count equals zero, the recombination frequency is very likely to be estimated as zero. Figure 3.2 shows the probability of zero *aa,BB* observations depending on the real recombination frequency for population sizes of 100 and 200. It shows, for instance, that for a reasonably sized F$_2$, i.e. 100 plants, there is a more than 35% chance of obtaining zero *aa,BB* plants and thus estimates of recombination frequency of zero. Even for an F$_2$ of double that size, the probability is more than 13%. It will be obvious that estimating reliable map orders of dominant markers in repulsion phase with each other is virtually impossible in practice.

3.4 Advanced population types

By advanced, we mean that the population is the result of having passed through multiple rounds of meiosis with segregation and recombination. Whilst, for instance, the F$_2$ is the result of segregation and recombination of the meiosis of the F$_1$ alone, the so-called *recombinant inbred lines* (RIL) family is the result of segregation and recombination of several generations of selfing using single-seed descent starting from the F$_2$ (*Remark:* with single-seed descent, you take a single seed from each plant as its offspring). Some other of these advanced population types are the *backcross inbred lines* (BIL) family, the *advanced intercross lines* (AIL) family and the family of F_2-derived doubled haploids (DH). These population types are always constructed with a specific goal in mind. For instance, RILs and DHs are created to obtain an immortalized population that may be utilized in multiple investigations. AILs are produced to achieve a high mapping resolution. Similarly, F_2-derived DHs have had an extra round of recombination over F_1-derived DHs. The use of BILs was proposed to enable a better way of studying (quantitative) traits within the genetic background of the backcross parent when trying to introduce traits from unadapted germplasm.

Normally in these advanced population types, linkage analysis is performed on the final generation only. It will then be unknown in which meiosis certain genotypes were formed. For instance, in the making of RILs, a genotype homozygous at two loci may be formed as the offspring of an F_2 individual and be propagated to the final RIL generation, or it may be created from an F_6 individual that was still present as a heterozygous genotype. By definition, the recombination frequency is the proportion of recombinations between two loci in a *single* meiosis. In its estimation procedure, it is assumed that the probability of recombination is the same over all subsequent meioses that resulted in the final population. In order to apply the method of maximum likelihood, the probabilities of each phenotype, P_p, must be expressed in terms of the recombination probability, r. Denoting the number of individuals of a given phenotype as n_p, then the likelihood is (ignoring the multinomial coefficient):

$$L = \prod_p (P_p)^{n_p},$$

and the log-likelihood is:

$$\ln(L) = \sum_p n_p \ln(P_p).$$

The log-likelihood is then maximized to obtain the estimate of the recombination frequency. For the F_2-derived DHs, the phenotype probabilities can be derived relatively easily:

Locus A	Locus B				MP
	BB		bb		
AA	AA,BB	$(2-3r+2r^2)/4$	AA,bb	$(3r-2r^2)/4$	$\frac{1}{2}$
aa	aa,BB	$(3r-2r^2)/4$	aa,bb	$(2-3r+2r^2)/4$	$\frac{1}{2}$
MP		$\frac{1}{2}$		$\frac{1}{2}$	

Similarly, the likelihood equation can be solved explicitly to obtain the estimator for r:

$$\hat{r} = \tfrac{3}{4} - \tfrac{1}{4}\sqrt{(9 - 16\,R/N)}.$$

For the other advanced population types, however, the numbers of generations of intercrossing, backcrossing and selfing are variable (i.e. depend on the experiment), making it difficult to derive general formulae for the phenotype probabilities in each specific situation. For the RIL family, complete sets of formulae with the number of selfings as a variable have been given by, for example, Bulmer (1985). Darvasi & Soller (1995) derived a formula for the case of the AIL family where there are no selfings following the intercrossing. In practice, however, there are usually some selfings in the construction of AILs (Winkler *et al.*, 2003). Because of the relative complexities of the phenotype probabilities for these population types, however, it is not easy to derive explicit maximum likelihood solutions. Therefore, the estimates of the recombination frequency are generally obtained using a standard numerical method. In this procedure, the phenotype probabilities for varying situations may be obtained in a relatively straightforward way: create a vector of probabilities for each phenotype and multiply the vector with a matrix defining the transition to the next generation. The multiplication is done repeatedly corresponding to the generation number of the specific population and if necessary with different transition matrices corresponding to the types of crossing the population has undergone.

Exercises

Please note, some of the exercises in this book use data files that are available to download at the Cambridge University Press web site at http://www.cambridge. org/9781107013216.

Exercise 3.1. Estimating recombination frequencies in a backcross
Of a backcross, F_1: (P_1 *AA* $\times$ P_2 *aa*) $\times$ P_1 *AA*, three markers, denoted by K, L and M, were scored on all 100 offspring individuals. The observations were recorded in the following way: *AA* denotes the genotype identical to P_1 and *Aa* denotes the genotype identical to the F_1. In the offspring, the following frequencies were determined:

K	L	M	Frequency	K	L	M	Frequency
AA	*AA*	*AA*	37	*Aa*	*AA*	*AA*	8
AA	*AA*	*Aa*	0	*Aa*	*AA*	*Aa*	5
AA	*Aa*	*AA*	4	*Aa*	*Aa*	*AA*	1
AA	*Aa*	*Aa*	5	*Aa*	*Aa*	*Aa*	40

Estimate the recombination frequencies between the markers K, L and M.

Exercise 3.2. Estimating recombination frequencies for a cross in apple

Below is the dataset of a mapping experiment in apple with the goal of mapping the scab resistance gene **D** (scab is caused by *Venturia inaequalis*). The progeny of 164 apple trees were scored for three markers, **A**, **B** and **C**, and the monogenic scab resistance. Although the progeny is a full-sib family of a cross between two outbred genotypes, the scab resistance and the three markers segregate as in a backcross: $Aa \times aa$. The genotypes of the markers (scored with *a* as homozygous and *h* as heterozygous) and the resistance (susceptible: *a*, resistant: *h*) are given. Observations that could not be made reliably are indicated as '−'.

Calculate all six pairwise recombination frequencies. The data are available in the file *Exercise_3.2.txt* as tab-delimited text, which can be imported or copied into spreadsheet software. Calculate the recombination frequency of at least the first pair without using spreadsheet software.

Tree	A	B	C	D	Tree	A	B	C	D	Tree	A	B	C	D
001	*h*	*h*	*h*	*h*	012	*a*	*a*	*h*	*a*	023	*a*	*a*	*a*	*a*
002	*a*	*a*	*h*	*a*	013	*h*	*h*	*a*	*h*	024	−	−	−	−
003	*h*	*h*	*a*	*h*	014	*h*	−	*h*	*h*	025	*a*	*a*	*a*	*a*
004	*h*	*h*	*h*	*h*	015	*a*	*a*	*a*	*a*	026	*a*	*a*	*a*	*a*
005	*h*	*h*	*a*	*h*	016	*h*	−	*a*	*a*	027	*h*	*h*	*h*	*h*
006	*h*	*a*	*h*	*a*	017	*h*	*h*	*h*	*h*	028	*a*	*a*	*a*	*a*
007	*h*	*h*	*a*	*h*	018	*h*	*a*	*h*	*h*	029	*h*	*h*	*h*	−
008	*h*	*h*	*h*	*h*	019	*a*	*a*	*a*	*a*	030	*a*	*a*	*a*	*a*
009	*h*	*h*	*h*	*h*	020	*a*	*h*	*h*	−	031	*a*	*a*	*a*	*a*
010	*h*	*a*	*h*	*a*	021	*a*	*a*	*a*	*a*	032	*a*	*a*	*a*	*a*
011	*a*	*a*	*a*	*a*	022	*h*	*h*	*h*	−	033	*a*	*a*	*a*	*a*

Tree	A	B	C	D	Tree	A	B	C	D	Tree	A	B	C	D
034	a	a	h	h	063	a	a	a	a	092	a	a	a	a
035	a	h	h	h	064	h	h	h	h	093	a	a	h	h
036	h	h	h	h	065	h	h	h	h	094	a	a	h	a
037	a	h	h	h	066	a	a	a	a	095	a	a	a	a
038	–	–	–	–	067	h	h	h	h	096	–	h	h	h
039	a	a	a	a	068	h	a	a	a	097	a	a	a	a
040	h	h	h	h	069	h	h	h	h	098	h	h	h	h
041	a	a	h	h	070	h	h	h	h	099	a	a	a	a
042	h	h	h	h	071	h	h	–	h	100	a	a	a	a
043	h	h	h	h	072	h	h	a	a	101	h	h	h	h
044	–	h	h	h	073	h	h	a	h	102	a	a	a	a
045	h	h	h	h	074	h	h	h	h	103	h	h	h	h
046	h	h	h	h	075	h	h	h	h	104	a	a	h	h
047	a	a	a	a	076	h	h	h	h	105	h	a	a	a
048	–	a	a	a	077	–	–	–	–	106	a	a	a	a
049	a	h	h	h	078	–	h	h		107	a	a	a	a
050	a	a	a	a	079	–	h	h	h	108		a	–	a
051	a	a	h	a	080	a	a	a	a	109	h	–	a	a
052	a	h	h	h	081	–	a	a	a	110	a	a	a	a
053	–	–	–	–	082	a	a	–	a	111	a	a	a	a
054	a	a	a	a	083	h	h	h	h	112	a	–	a	a
055	h	h	h	h	084	a	h	a	h	113	–	–	–	–
056	a	a	a	a	085	a	a	a	a	114	h	–	a	a
057	–	h	h	–	086	–	h	h	–	115	a	a	a	a
058	a	a	h	a	087	a	a	a	a	116	a	a	a	a
059	–	h	a	–	088	h	h	h	h	117	a	a	–	a
060	h	h	h	h	089	h	h	a	a	118	a	a	a	a
061	a	a	h	a	090	a	a	a	a	119	a	a	a	a
062	h	h	h	h	091	a	a	a	a	120	a	a	a	a

Tree	A	B	C	D	Tree	A	B	C	D	Tree	A	B	C	D
121	a	h	h	h	136	a	h	h	h	151	a	a	a	a
122	h	h	h	h	137	a	a	a	a	152	h	h	h	h
123	a	h	h	h	138	–	a	a	a	153	a	a	a	a
124	–	h	h	h	139	h	h	h	h	154	a	a	a	a
125	a	h	h	h	140	h	h	h	h	155	a	a	a	a
126	h	h	h	h	141	h	h	h	h	156	h	h	h	h
127	–	h	h	–	142	h	a	a	a	157	a	a	a	a
128	a	a	a	a	143	h	h	a	h	158	h	h	h	h
129	–	a	a	a	144	–	–	–	–	159	h	h	–	h
130	–	a	a	a	145	a	a	a	a	160	a	a	a	a
131	h	h	h	h	146	h	h	h	h	161	h	–	h	h
132	–	h	–	h	147	a	a	a	a	162	a	a	a	a
133	a	a	a	a	148	h	h	a	a	163	h	h	h	h
134	a	h	h	h	149	–	a	h	h	164	a	h	–	h
135	h	h	h	h	150	a	h	a	a					

Exercise 3.3. Estimating recombination frequencies in an F_2

In an F_2 of 100 individuals, the following phenotypes for loci **A** and **B** were observed:

	Locus B		
Locus A	*B*	*Bb*	*bb*
AA	18	4	1
Aa	5	44	1
aa	0	4	23

A. Suppose that $r = 0.25$ (starting value); calculate the expected proportion of individuals in repulsion phase for the class of double heterozygotes.
B. Calculate the expected number of recombinations for the class of double heterozygotes, supposing $r = 0.25$.
C. Calculate the total number of recombinations in the other eight classes.

D. Calculate the expected total number of recombinations for all classes together, supposing that $r = 0.25$.
E. Obtain a new value for the recombination frequency.
F. Repeat questions **(A)** to **(E)** with the value of the recombination frequency obtained in **(E)** instead of 0.25. (*Hint: try to program all above steps in a spreadsheet. Save the spreadsheet file for the next exercise.*)
G. Repeat questions **(A)** to **(E)** with the value of the recombination frequency obtained in **(F)** instead of 0.25.

Exercise 3.4. Estimating recombination frequencies in an F_2 with dominant markers in coupling and repulsion phase

An F_2 of 100 individuals was simulated with four markers, **m1**, **m2**, **m3** and **m4**, on one chromosome with a recombination frequency of $r = 0.13$ (=15 cM map distance) between neighbours. The map order was **m1–m2–m3–m4**. The markers were scored in a codominant fashion (*AA* like P_1, *Aa* like F_1 and *aa* like P_2) and the following 3×3 contingency tables of the phenotype frequencies of the six pairs were obtained:

	m2				**m3**				**m4**		
m1	*AA*	*Aa*	*aa*	**m1**	*AA*	*Aa*	*aa*	**m1**	*AA*	*Aa*	*aa*
AA	18	11	0	*AA*	18	8	3	*AA*	15	11	3
Aa	6	3	8	*Aa*	10	27	14	*Aa*	12	24	15
aa	0	4	16	*aa*	0	6	14	*aa*	3	10	7

	m3				**m4**				**m4**		
m2	*AA*	*Aa*	*aa*	**m2**	*AA*	*Aa*	*aa*	**m3**	*AA*	*Aa*	*aa*
AA	19	5	0	*AA*	16	8	0	*AA*	23	5	0
Aa	9	32	11	*Aa*	12	27	13	*Aa*	5	33	3
aa	0	4	20	*aa*	2	10	12	*aa*	2	7	22

A. Estimate the recombination frequency for all six pairs. (*Hint: use the spreadsheet developed in the preceding exercise.*)
B. Calculate the *LOD*s for all pairs.
C. Suppose the *a* allele of P_2 was dominant over the *A* allele of P_1 for all four markers. What would become of the contingency tables in this case?
D. Estimate the six recombination frequencies and *LOD*s for this case. Compare the results with the codominant situation.

E.   Suppose the *a* allele of P_2 was dominant over the *A* allele of P_1 for markers **m1** and **m3**, whereas the *A* allele of P_1 was dominant over the *a* allele of P_2 for markers **m2** and **m4**. What would become of the contingency tables in this case?

F.   Estimate the six recombination frequencies and *LOD*s for this case. Compare the results with the codominant situation.

References

Bateson, W. (1909). *Mendel's Principles of Heredity*. Cambridge: Cambridge University Press. http://www.archive.org.

Bulmer, M. G. (1985). *The Mathematical Theory of Quantitative Genetics*. Oxford: Clarendon Press.

Darvasi, A. & Soller, M. (1995). Advanced intercross lines, an experimental population for fine genetic mapping. *Genetics*, 141, 1199–207.

Dempster, A. P., Laird, N. M. & Rubin, D. B. (1977). Maximum likelihood from incomplete data via the EM algorithm. *Journal of the Royal Statistical Society Series B*, 39, 1–38.

Johannsen, W. (1909). *Elemente der exakten Erblichkeitslehre*. Jena: Gustav Fischer Verlag. http://www.archive.org.

Mather, K. (1936). Types of linkage data and their value. *Annals of Eugenics*, 7, 251–64.

Mather, K. (1951). *The Measurement of Linkage in Heredity*. London: Methuen.

Punnett, R. C. (1905). *Mendelism*. Cambridge: Macmillan & Bowes.

Sturtevant, A. H. (1913). The linear arrangement of six sex-linked factors in *Drosophila*, as shown by their mode of association. *Journal of Experimental Zoology*, 14, 43–59. http://www.esp.org/timeline.

Winkler, C. R., Jensen, N. M., Cooper, M., Podlich, D. W. & Smith, O. S. (2003). On the determination of recombination rates in intermated recombinant inbred populations. *Genetics*, 164, 741–5.

4 Determination of linkage groups

Genes and markers on the same chromosome have a linear arrangement, which can be described by a (linear) map. Genes and markers on different, non-homologous chromosomes are inherited independently and therefore there is no linear arrangement between them. This is the reason why genetic linkage maps are estimated separately for each chromosome. Determining which genes and markers belong to the same chromosome is therefore a necessary preparation for map construction. Sets of linked loci are called linkage groups. Ideally, the number of linkage groups is the same as the haploid number of chromosomes. In practice, this is not always the case. Sometimes, the set of loci studied does not cover the entire genome or is not distributed evenly across the genome. On other occasions, spurious linkage causes loci from separate chromosomes to end up in a single linkage group, which is a more common problem to solve.

4.1 Why determine linkage groups and how?

Chromosomes are very large linear molecules. Genes and genetic markers are, or represent, tiny parts of the chromosomes. They are referred to as loci. Their linear arrangement on the chromosomes can be described with mutual distance measures, for instance in base pairs or in recombination units. A linkage map is such a description. Because a linear arrangement exists only for loci of the same chromosome, determining which loci reside on the same chromosome is the first step in the construction of a linkage map. Loci on the same chromosome are physically linked, whereas those on different (non-homologous) chromosomes are physically unlinked. Genetic linkage is the phenomenon whereby traits have a tendency to be inherited together. Due to independent assortment of the chromosomes in meiosis, genetic linkage is the result of physical linkage. Determining whether two loci are genetically linked is the starting point of lumping loci into groups. It is preferable to identify such groups of linked genes and markers as *linkage groups* rather than as chromosomes, as long as the physical relationship between the loci and the chromosomes is not established (e.g. through cytogenetics). This is even more relevant if there are more groups than homologous chromosome pairs. Such situations often occur in experiments where the genome is not covered completely with markers.

Deciding whether two loci are genetically linked can be answered by carrying out a statistical test. The basis of this test is the meiotic property of the independent assortment of the chromosomes. Because of this, the alleles of loci residing on different chromosomes segregate independently. In contrast, the alleles of loci on the same chromosome pair segregate more often in the same (parental) combinations or, in statistical terms: the alleles of these two loci are *associated* in their segregation. In other words, the loci are genetically linked. Derived from the independent assortment of chromosomes is the property that the recombination frequency is equal to $\frac{1}{2}$. There are two principal ways to test for linkage. These start from the following hypotheses:

1. Alleles of loci segregate independently.
2. The recombination frequency is equal to $\frac{1}{2}$.

The tests are carried out on all pairs of loci. If, according to the test, a pair of loci is unlinked, the pair may still be linked but indirectly through other pairs. For instance, if locus **A** is significantly linked to locus **B** but not to locus **C**, whilst locus **B** is linked to locus **C**, then locus **A** is indirectly linked to locus C. All loci that are linked, either directly or indirectly, will form a linkage group.

4.2 Test for independent segregation

The primary characteristic of meiosis regarding the relationship of loci on different chromosomes is their independent segregation. The standard statistical test of whether two loci segregate independently is the test for independence in a contingency table. In this test, observed frequencies of genotypes are compared with their expected frequencies under the assumption of independence using a χ^2 statistic or a likelihood ratio statistic. We will start with the backcross and generalize later. The backcross $(AA,BB \times aa,bb) \times aa,bb$ will have in the segregating population two genotypes for locus **A**, Aa and aa, and for locus **B**, Bb and bb. For the pair of loci **A** and **B**, the individuals of a backcross can be classified according to their two-locus phenotype ($=$ genotype in the backcross) into a 2×2 contingency table, where the rows refer to the phenotypes of the first locus and the columns to the phenotypes of the second locus. The table shows the observed frequencies of each two-locus phenotype, n_p; the row and column totals t_p and the overall total N are added to the table:

Locus A	Locus B		Total	(4.1)
	Bb	bb		
Aa	$n_{Aa,Bb}$	$n_{Aa,bb}$	t_{Aa}	
aa	$n_{aa,Bb}$	$n_{aa,bb}$	t_{aa}	
Total	t_{Bb}	t_{bb}	N	

Under the hypothesis that locus **A** segregates independently from locus **B**, the probability of each of the four cells in the table is equal to the product of the corresponding row and column probabilities. These row and column probabilities can be estimated by dividing the row and column totals by the overall total. Estimates of the expected frequencies of individuals of the four cells of the table are obtained by multiplying the cell probabilities with the overall total. For instance, for Aa,Bb, the estimate of the expected number of individuals becomes:

$$e_{Aa,Bb} = (t_{Aa}/N)(t_{Bb}/N)\,N = t_{Aa}\,t_{Bb}/N.$$

The expected frequencies can also be arranged in a 2×2 contingency table:

Locus A	Locus B		Total	(4.2)
	Bb	bb		
Aa	$e_{Aa,Bb}$	$e_{Aa,bb}$	t_{Aa}	
aa	$e_{aa,Bb}$	$e_{aa,bb}$	t_{aa}	
Total	t_{Bb}	t_{bb}	N	

It should be noted that the marginal totals, t_p, of (4.2) and (4.1) are the same. There are two standard statistics for testing the hypothesis of independent segregation, the Pearson χ^2 statistic:

$$X^2 = \sum_p (n_p - e_p)^2/e_p,$$

and the likelihood ratio χ^2 statistic:

$$G^2 = 2\sum_p n_p(\ln(n_p) - \ln(e_p)),$$

in which $\ln()$ is the natural logarithm function and the sums (Σ) are taken over all four phenotype cells of the table. Both statistics behave approximately according to a χ^2 distribution with one degree of freedom. For large values of n_p, the two statistics are equivalent. The results of the test should be considered with care if one or more of the cell expectations is smaller than five. Deviations from independent segregation will lead to large values of the test statistics.

Example

Suppose our backcross population consists of 100 individuals, in which we find 41 individuals with genotype *Aa,Bb*, 39 with genotype *aa,bb*, 8 with genotype *Aa,bb* and 12 with genotype *aa,Bb*. Then the 2×2 contingency table for these two loci becomes:

(4.3)

Locus A	Locus B		Total
	Bb	*bb*	
Aa	41	8	49
aa	12	39	51
Total	53	47	100

The expected frequencies under the assumption of independent segregation become:

Locus A	Locus B		Total
	Bb	*bb*	
Aa	$49 \times 53/100 = 25.97$	$49 \times 47/100 = 23.03$	49
aa	$51 \times 53/100 = 27.03$	$51 \times 47/100 = 23.97$	51
Total	53	47	100

Now we can calculate the Pearson χ^2:

$$X^2 = (41 - 25.97)^2/25.97 + (8 - 23.03)^2/23.03 +$$
$$(12 - 27.03)^2/27.03 + (39 - 23.97)^2/23.97 = 36.29,$$

and the likelihood ratio χ^2:

$$G^2 = 2\,[\,41(\ln(41) - \ln(25.97)) + 8\,(\ln(8) - \ln(25.97)) +$$
$$12\,(\ln(12) - \ln(27.03)) + 39\,(\ln(39) - \ln(23.97))] = 39.00\,.$$

As the 95% threshold value of the χ^2 distribution with one degree of freedom is 3.84, both statistics are highly significant, meaning that the loci do not segregate independently.

Generalization

If the segregating population type generates more phenotypes, then the contingency table takes on a larger dimension. For instance, in an F_2 with codominant loci, which have three phenotypes per locus, the table will be 3×3 in size. The estimation of expected frequencies is similar to that in the 2×2 table, i.e. the product of the corresponding estimated row and column probabilities and the overall total of individuals. The calculation of the two test statistics is a straightforward extension, i.e. the sum is taken over all phenotype cells; thus, in this F_2 example, over nine cells. The number of degrees of freedom is equal to the number of rows minus one, multiplied by the number of columns minus one; thus, in this F_2 example: $(3 - 1)(3 - 1) = 4$.

4.3 Test for the recombination frequency being equal to $\frac{1}{2}$

A derived characteristic of meiosis regarding the relationship between loci on different homologues is that their recombination frequency is expected to be $\frac{1}{2}$. The test for the hypothesis that the recombination frequency is equal to $\frac{1}{2}$ is an alternative test for linkage. The standard statistic for this test is the deviance, D, which is twice the logarithm of the likelihood ratio. The likelihood ratio is the quotient of two likelihoods of two different statistical models (L_1 and L_0). Thus, the deviance is:

$$D = 2\ln(L_1/L_0) = 2(\ln(L_1) - \ln(L_0)).$$

Here, L_0 represents the likelihood of the null hypothesis that the recombination frequency equals $\frac{1}{2}$, whilst L_1 represents the likelihood of the alternative hypothesis that the recombination frequency equals the maximum likelihood estimate ($r = \hat{r}$). Under the null hypothesis, the deviance follows a χ^2 distribution with one degree of freedom. In the previous chapter, the log-likelihood was derived for the backcross:

$$\ell = \ln(L(r)) \propto (N - R)\ln(1 - r) + R\ln(r) = S\ln(s) + R\ln(r),$$

with R being the total number of recombinant individuals observed: $n_{Aa,bb} + n_{aa,Bb}$, $S = N - R$, the non-recombinant individuals, and $s = 1 - r$. The maximum likelihood estimator for the recombination frequency was:

$$\hat{r} = R/N.$$

Thus, the deviance for the backcross becomes:

$$D = 2\left[S\ln(\hat{s}) + R\ln(\hat{r}) - S\ln\left(\frac{1}{2}\right) - R\ln\left(\frac{1}{2}\right)\right]$$
$$= 2[S\ln(\hat{s}) + R\ln(\hat{r}) + N\ln(2)].$$

In genetics, normally a derived test statistic is used, which uses the common logarithm (i.e. the logarithm to base 10, $\log_{10}$) instead of the natural logarithm (i.e. the logarithm to base $e \approx 2.718$: ln or $\log_e$). The statistic is called the logarithm of odds or the *LOD*:

$$LOD = \log_{10}(L_1/L_0) = \log_{10}(L_1) - \log_{10}(L_0).$$

Note that the *LOD* does not have the factor 2 present in the formula as in the deviance formula. Because the following mathematical relationship holds between the natural and the common logarithm:

$$\ln(x) = \ln(10)\log_{10}(x),$$

it follows that the *LOD* multiplied by $2\ln(10) \approx 4.605$ is equal to the deviance, which enables testing against the χ^2 distribution. For the backcross, the *LOD* becomes:

$$LOD = S\log_{10}(\hat{s}) + R\log_{10}(\hat{r}) - S\log_{10}\left(\tfrac{1}{2}\right) - R\log_{10}\left(\tfrac{1}{2}\right)$$
$$= S\log_{10}(\hat{s}) + R\log_{10}(\hat{r}) + N\log_{10}(2).$$

Example

Let us continue with the backcross example (4.3) and apply this test. The four two-locus genotypes can be grouped into a recombinant (*Aa,bb* and *aa,Bb*) and a non-recombinant (*Aa,Bb* and *aa,bb*) class. The number of individuals in the recombinant class is $R = 8 + 12 = 20$ and that in the non-recombinant class is $S = 41 + 39 = 80$; the total is $N = 100$. The maximum likelihood estimate of the recombination frequency is $\hat{r} = R/N = 20/100 = 0.2$. Hence, the deviance becomes:

$$D = 2\,[80\ln(0.8) + 20\ln(0.2) + 100\ln(2)] = 38.55\,,$$

and the *LOD* becomes:

$$LOD = 80\log_{10}(0.8) + 20\log_{10}(0.2) + 100\log_{10}(2) = 8.37.$$

You may verify with these results that indeed: $D/LOD = 2\ln(10)$. When compared with the 95% threshold value of the χ^2 distribution with one degree of freedom, 3.84, the deviance is highly significant, meaning that the recombination frequency between the loci is significantly different from $\tfrac{1}{2}$.

Generalization

As we have seen in the previous chapter, different population types have different formulae for the likelihoods and different estimators for the recombination frequency. However, for all situations, the general formulae for the deviance and *LOD* remain applicable:

$$D = 2\ln(L_1/L_0)$$

and

$$LOD = \log_{10}(L_1/L_0).$$

In all cases, the degree of freedom of the χ^2 distribution, against which it is tested, is one.

4.4 Some caution

The different tests for linkage of the two previous sections as applied to the backcross example (4.3) lead to similar χ^2 values: the Pearson χ^2 was 36.29 and the deviance was 38.55. This is reassuring, as the tests are based on different model assumptions.

Nature, however, does not always behave according to our models. A common phenomenon is the occurrence of (natural) *selection* of gametes and zygotes. As a result, certain alleles or even allele combinations of multiple loci are favoured over others in the offspring. This leads to the observation of so-called *segregation distortion*: the frequencies of locus genotypes are not according to the expected Mendelian frequencies but are skewed, and sometimes much skewed, towards one or the other genotype. If the selection is performed simultaneously on two or more loci residing on different non-homologous chromosomes, then the resulting relationship between the loci in the offspring is similar to linkage. To illustrate the problem, we will use the following numerical example of the backcross:

	Locus B		
Locus A	*Bb*	*bb*	**Total**
Aa	1	9	10
aa	11	79	90
Total	12	88	100

Here, at both loci, the genotype with the allele from the second parent (lower case) is present more frequently (0.90, 0.88) than the genotype carrying the allele from the first parent (upper case) (0.10, 0.12). The segregation for each locus is quite different from the Mendelian 1:1 expectation for the backcross. The numbers of recombinant ($R = 9 + 11 = 20$) and non-recombinant individuals ($S = 80$)

are the same as in the example of the previous section, and as a consequence the estimate of the recombination frequency and the *LOD* are the same: $r = 0.2$, $LOD = 8.37$ (and $D = 38.55$). So the conclusion would be as in the previous section: loci **A** and **B** are linked. However, if we apply the tests for independence, the results are contradictory. The expected frequencies under the assumption of independent segregation are:

	Locus B	
Locus A	*Bb*	*bb*
Aa	$10 \times 12/100 = 1.2$	$10 \times 88/100 = 8.8$
aa	$90 \times 12/100 = 10.8$	$90 \times 88/100 = 79.2$

With these values, we can calculate the two test statistics: $X^2 = 0.042$ and $G^2 = 0.044$ (on the *LOD* scale, 0.009 and 0.010), respectively. These results are certainly not significant (the mean of a χ^2 variable with one degree of freedom is equal to 1!). In other words, according to both tests for independent segregation, the loci **A** and **B** are unlinked.

This latter result is in clear contrast to the conclusion of the test for $r = \frac{1}{2}$. Under segregation distortion, this test for $r = \frac{1}{2}$ cannot distinguish between a situation of linked loci and a situation of loci on different chromosomes. Thus, this test may lead to false-positive conclusions regarding linkage in these cases. The test for independence, however, loses power under segregation distortion, irrespective of whether the loci are on the same chromosome or not. Therefore, this test does not lead to false-positive (spurious) linkages.

Due to false-positive linkages, the map estimation will attempt to make a single map of markers on two or more chromosomes, which is usually problematic. Therefore, it is preferable to prevent false-positive linkage situations. Consequently, when forming linkage groups using the test for $r = \frac{1}{2}$, loci with (severe) linkage distortion should be ignored.

The independence test has its own drawback. If there is a significant deviation from independence, then this does not necessarily mean that the deviation is due to linkage: the data may have been coded in the wrong way, or there may be other serious errors in the data.

4.5 Grouping algorithm

After deciding whether loci are linked or not, the determination of a linkage group is more or less straightforward. A simple algorithm could read:

1. Assess all pairs of loci to find two linked loci that are not yet placed in a group; assign these loci to a new group.
2. Subsequently, assess all pairs of loci and if a locus is linked to a locus that is a member of the group, then add the locus to the group.
3. Repeat step (2) until no more loci are added to the group.
4. Continue with step (1) and start with a new group, until either all loci are grouped or no more new groups are created.

However, this procedure has the drawback that it has to loop repeatedly over many pairs of loci, whilst the number of pairs may be huge: it increases quadratically with the number of loci. Thus, in situations with many loci, the above procedure will be very time consuming. Therefore, a faster approach, in which all pairs have to be assessed only once, is the following:

1. Assess the next pair of loci (start at the first pair).
2. If the pair is unlinked, then continue with step (1).
3. If the pair is linked, then one of four situations occurs:
    (a) Neither loci have been assigned to a group, so assign them to a new group.
    (b) One of the loci has been assigned to a group, so assign the other locus to this group.
    (c) Both loci have been assigned to different groups, so merge the groups.
    (d) Both loci have been assigned to the same group, so take no further action.
4. Continue with step (1), until either all loci are grouped or all pairs are dealt with.

For a BC_1 and a DH_1, the recombination frequency is equal to one minus the simple matching similarity coefficient. This coefficient is often used in cluster analysis and principal coordinate analysis in connection with binary data. The grouping of loci may be considered as a special form of hierarchical cluster analysis using the single linkage criterion. An efficient algorithm for forming groups from a matrix of links is known as a depth-first search (Cormen *et al.*, 2001).

4.6 Determining the linkage groups in practice

The ultimate goal of the grouping procedure is to end up with as many linkage groups as there are homologous chromosome pairs. If possible, cytogenetic evidence should provide the relationship between linkage groups and chromosomes. The theory is straightforward; the practice, however, is often fraught with difficulties.

The first problem that may occur is an uneven coverage of the genome by the set of genotyped loci. Usually, a mapping project starts with a certain number of anonymous markers. These are assumed to be spread evenly over the genome. In reality, this may not be the case: large stretches of the genome may not be covered. The method with which molecular markers are acquired may influence the coverage. For instance, markers may be obtained selectively from coding regions (or, inversely, from non-coding regions), and these regions are not necessarily spread evenly across the genome. The result of having longer stretches of uncovered genome will be that there will not be (significant) linkage across uncovered stretches. Consequently, there will be more linkage groups than chromosome pairs. This problem can only be solved by filling the uncovered stretches of genome with additional markers.

Another problem is which significance threshold to apply. Simply using a 5% significance level on every tested pair of loci will result in loci from different chromosomes being declared linked simply by chance: because many pairs are tested, there is a high chance of false positives. Originally in linkage studies, as a rule of thumb often a *LOD* threshold of 3 was used, (i.e. linkage is $10^3 = 1000$ times more likely). This corresponds to a χ^2 value of 13.81 with a *P* value of 0.0002 (i.e. 0.02%), which is much more stringent than the 5% of a single test. Current experience shows that this level is often insufficiently stringent to obtain the number of linkage groups that corresponds to the number of homologous chromosome pairs. Apparently, there is in practice a certain level of spurious (false-positive) linkage. Presumably, this is caused by some form of selection affecting multiple chromosomes. (*Remark*: although hopefully not, it might also be caused by gross errors in the genotype observations. Because mapping experiments involve so many steps and individuals, from creating the cross and growing the individuals through all the laboratory actions until generating the data files that are ready for analysis, it is necessary to be really meticulous in all steps in order to prevent errors slipping in somewhere in the entire procedure.)

Because of this difficulty, it is advisable to start with a grouping that has more linkage groups than chromosome pairs. Next, estimate the maps for these groups and subsequently use a grouping at a lesser stringency to see which higher-stringency groups possibly belong together. If the resulting map is more or less a fusion of the maps of the underlying groups with a big empty stretch in between, then the underlying groups are likely to belong together. However, if the resulting map mixes the orders of the maps of the underlying groups, then the underlying groups do not belong to the same chromosome.

A tool that can assist here is a display of the linkage groups in a tree structure, where the nodes represent linkage groups at a certain test threshold and the branches

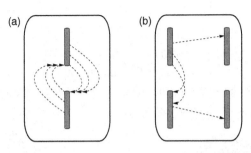

Fig. 4.1

Checking the strongest cross-links (arrows) may reveal spurious linkage. (a) Subgroups for a chromosome with a stretch without loci: all the links point in the same direction. (b) Unrelated groups displaying spurious linkage: the links point in different directions to different groups.

represent transitions to a test threshold with a different stringency. Nodes with more than a single branch in the direction of increasing stringency indicate that the group splits into multiple linkage groups at that level of stringency. This structure makes it easy to find groups that possibly belong to each other at a lesser stringency, as described in the previous paragraph.

Linkage groups are formed simply by adding markers that have significant linkage to one marker already in the group. Spurious linkage between just two markers from two different chromosomes will result in the joining of the groups they are in. A diagnostic tool that may provide insight into the relationship between loci is principal coordinate analysis (Chatfield & Collins, 1980; Cox & Cox, 2000). A chart of the scores of the loci on the first three principal axes may visualize which markers are responsible for linking up loci from different chromosomes. An example of this use of principal coordinate analysis is given by Farré *et al.* (2011). These authors investigated the consequences to linkage analysis of a reciprocal translocation in one of the parents of a cross in barley. Because recombination does not occur at the translocation breakpoints, this leads to a complicated linkage relationship between the loci on the chromosomes involved in the translocation.

A simpler tool for this purpose provides the determination for every locus of the so-called *strongest cross-link* values. This is defined as the strongest linkage value (e.g. r, LOD, X^2) found for a locus in a group with any other locus outside its own group. Once the separate maps are calculated, the strongest cross-links should all point in the same direction if the subgroups really represent two parts of the same chromosome (Fig. 4.1a). In unrelated subgroups that appear to be linked due to spurious linkage, the pattern of the strongest cross-links will be more chaotic (Fig. 4.1b).

Exercises

Exercise 4.1. Testing linkage between two markers in a backcross

In a normal backcross population (BC_1), the genotypes of markers **A** and **B** were observed for 100 plants. From the data, the following contingency table for the combined genotypes was constructed:

Locus A	Locus B	
	Bb	*bb*
Aa	32	17
aa	10	41

A. Estimate the recombination frequency.
B. Calculate the *LOD* value.
C. Calculate the X^2 statistic.
D. Calculate the G^2 statistic.
E. Are the markers linked?

Exercise 4.2. Testing linkage between two markers in a backcross

In a normal backcross population (BC_1), the genotypes of markers **A** and **B** were observed for 100 plants. From the data, the following contingency table for the combined genotypes was constructed:

Locus A	Locus B	
	Bb	*bb*
Aa	25	50
aa	0	25

A. Estimate the recombination frequency.
B. Calculate the *LOD* value.
C. Calculate the X^2 statistic.
D. Calculate the G^2 statistic.
E. Are the markers linked?

Exercise 4.3. Testing linkage between a codominant and a dominant marker in an F_2

In a normal F_2 population, the genotypes of markers **A** and **B** were observed for 100 plants. For marker **B**, the *b* allele was dominant over the *B* allele, so the *b* phenotype class comprises *Bb* and *bb* genotypes. From the data, the following contingency table for the combined phenotypes was constructed:

Locus A	Locus B	
	BB	*b–*
AA	18	11
Aa	8	44
aa	1	18

A. The maximum likelihood estimate of the recombination frequency is 0.2093. Calculate the *LOD* value.
B. Calculate the X^2 statistic.
C. Calculate the G^2 statistic.
D. How many degrees of freedom do the X^2 and the G^2 statistic have?
E. Are the markers linked?

References

Chatfield, C. & Collins, A. J. (1980). *Introduction to Multivariate Analysis*. London: Chapman & Hall.

Cormen, T. H., Leiserson, C. E., Rivest, R. L. & Stein, C. (2009) *Introduction to Algorithms*, 3rd edn. Cambridge, USA: MIT Press and McGrawHill.

Cox, T. F. & Cox, M. A. A. (2000). *Multidimensional Scaling*, 2nd edn. London: Chapman & Hall.

Farré, A., Lacasa Benito, I., Cistué, L., De Jong, J. H., Romagosa, I. & Jansen, J. (2011). Linkage map construction involving a reciprocal translocation. *Theoretical and Applied Genetics*, 122, 1029–37.

5 Estimation of a genetic map

The central idea of linkage analysis is that the rate of recombination between two loci is a reflection of their mutual distance on the chromosomes. Combining recombination frequencies between multiple loci on the same chromosome allows determination of the relative positions of these loci on the chromosome, thus creating a linkage map. A linkage map is defined by the order of its loci and by their mutual distances. This chapter is about obtaining the map distances for a given order.

5.1 Recombination frequencies and distances

One of the functions of a linkage map is to allow easy calculation of recombination frequencies between loci from their positions on the map. The linearity of a linkage map implies that distances can be added or subtracted. The question is whether recombination frequencies can be used as map distances. This problem was initially met by Sturtevant (1913), who was the first person to produce a genetic linkage map. At the time, it was becoming clear that genetic factors were located on the chromosomes. Sturtevant realized that it should be possible to determine the spatial arrangement of the genetic factors on the linear structure of a chromosome using recombination frequencies between factors as measures of their mutual distances. However, he found a discrepancy between the sum of adjacent distances and the direct measurement of the overall distance. He claimed that shorter distances were measured more accurately. He attributed the discrepancy to double recombinations and to a phenomenon now known as crossover interference. The essence of the problem is that recombination frequencies are not additive. The problem was solved by Haldane (1919), who derived a function that translates recombination frequencies into additive distances.

Suppose a backcross segregates for three ordered loci, **X**, **Y** and **Z**. The question is, how can we describe the mathematical relationship between the three pairwise recombination frequencies: r_{XY}, r_{YZ} and r_{XZ}? We have seen that the theoretical maximum value of the recombination frequency is 0.5. Suppose the three loci lie far apart, then r_{XY} and r_{YZ} would each be close to 0.5. Simply adding these values to obtain r_{XZ} would result in a value much more than 0.5, which is above the theoretical maximum. Adding recombination frequencies over more than two adjacent segments might even result in values greater than 1. This is impossible, because recombination frequencies are probabilities and probabilities can never be greater than 1. Thus, simply adding recombination frequencies of adjacent segments to obtain the recombination frequency over the joint segment is incorrect. The question is, what should r_{XZ} be in terms of r_{XY} and r_{YZ}, given that the order is **X–Y–Z**? To answer this question, we must know the following: when can we observe recombination between **X** and **Z**, considering whether or not there is recombination in the segments **X−Y** and **Y−Z**? There are four possible situations:

1. If there is *no* recombination in both segments, **X−Y** and **Y−Z**, then there is no recombination between **X** and **Z**.
2. If there *is* recombination in *both* segments, **X−Y** and **Y−Z**, then we will observe no recombination between **X** and **Z**.
3. Alternatively, if there *is* recombination between **X** and **Y**, whilst there is *no* recombination between **Y** and **Z**, then we will observe recombination between **X** and **Z**.
4. The same holds for the reciprocal situation: no recombination between **X** and **Y**, and recombination between **Y** and **Z**.

If we assume that recombination events in adjacent segments are independent, then probabilities of separate events in the adjacent segments can be multiplied to obtain the probability of combined events. Thus, we obtain the following relationship:

$$r_{XZ} = r_{XY}(1 - r_{YZ}) + (1 - r_{XY})r_{YZ} = r_{XY} + r_{YZ} - 2r_{XY}r_{YZ}. \qquad (5.1)$$

In other words, there is recombination between **X** and **Z** if there is recombination in the constituent segments **X−Y** or **Y−Z**, except when there is recombination in both. Because double recombinants are present in the class with recombination between **X** and **Y**, as well as in the class with recombination between **Y** and **Z**, the corresponding probability $r_{XY}\,r_{YZ}$ has to be subtracted twice from the sum $r_{XY} + r_{YZ}$. This formula shows that (under the independence assumption) the simple addition of two adjacent recombination frequencies needs a correction, which is equal to twice their product. It is important to realize that, if we are dealing with short segments, the values of r will be small, and therefore the term $2\,r_{XY}\,r_{YZ}$ will be very small and thus the discrepancy may be ignored; hence Sturtevant's claim that shorter distances were more accurate.

Similar formulae can be derived for situations with three or more neighbouring segments, but you can imagine that these quickly become intractable. In order to make things easier, Haldane (1919) derived a function that translates recombination frequencies, r, into map distances, d, which do have the property that they are additive:

$$d(r_{XZ}) = d(r_{XY}) + d(r_{YZ}).\tag{5.2}$$

This function is called a map or mapping function. Map distances are usually given in *centiMorgan* (cM) units, in honour of the early 20th-century geneticist T. H. Morgan who was one of the first to study the relationship (*Remark*: Sturtevant was one of his students). Haldane's mapping function for the relationship between r as the recombination frequency and d, the corresponding distance in cM, is:

$$d = -50 \ln(1 - 2r)\tag{5.3}$$

and its inverse function is (where $\exp(x)$ is the exponential function, e^x):

$$r = (1 - \exp(-0.02\,d))/2.$$

The Haldane mapping function is based on the fact that recombination events in neighbouring map segments are mutually independent, i.e. (5.1) holds. As an example, suppose:

$$r_{XY} = 0.05 \quad \text{and} \quad r_{YZ} = 0.10,$$

then:

$$r_{XZ} = 0.05 + 0.10 - (2 \times 0.05 \times 0.10) = 0.14,$$

whilst the translation into distances is:

$$d_{XY} = 5.268\,\text{cM} \quad \text{and} \quad d_{YZ} = 11.157\,\text{cM},$$

then:

$$d_{XZ} = 5.268 + 11.157 = 16.425\,\text{cM}.$$

The value 16.425 cM translates back to $r_{XZ} = 0.14$:

$$r_{XZ} = (1 - \exp(-0.02 \times 16.425))/2 = 0.14,$$

which is the same as in the direct calculation. With some straightforward mathematics, we can show that the Haldane mapping function (5.2) is equivalent to (5.1). By applying the mapping function (5.3) to the distances in (5.2), we derive:

$$-50 \ln(1 - 2r_{XZ}) = -50 \ln(1 - 2r_{XY}) + -50 \ln(1 - 2r_{YZ})$$
$$\Leftrightarrow 1 - 2r_{XY} = (1 - 2r_{XY})(1 - 2r_{YZ})$$
$$\Leftrightarrow 1 - 2r_{XZ} = 1 - 2r_{XY} - 2r_{YZ} + 4r_{ZY}r_{YZ}$$
$$\Leftrightarrow r_{XZ} = r_{XY} + r_{YZ} - 2r_{XY}r_{YZ},$$

which is (5.1). A final but practical remark is that if r is small, then the first-order Taylor series approximation of (5.3) will show that $d \approx 100\, r$ cM.

Results from quite large experiments had shown discrepancies from predictions based on (5.1). Apparently, the presence of one crossover had an effect on the occurrence of crossovers in neighbouring chromosome segments. One crossover may suppress the occurrence of a second crossover in a nearby segment, but the opposite situation may also occur. This phenomenon is called *crossover interference*. In 1944, Kosambi introduced a mapping function that described many experimental results much better than Haldane's mapping function. His mapping function reads:

$$d = 25 \ln((1 + 2r)/(1 - 2r))$$

and the inverse function is (where $\tanh(x)$ is the hyperbolic tangent, $\tanh(x) = (e^{2x} - 1)/(e^{2x} + 1)$):

$$r = 0.5 \tanh(0.02\, d).$$

Also, for this situation, an elegant relationship between the three recombination frequencies, r_{XY}, r_{YZ} and r_{XZ}, holds:

$$r_{XZ} = (r_{XY} + r_{YZ})/(1 + 4\, r_{XY}\, r_{YZ}).$$

Applying this to the example:

$$r_{XY} = 0.05 \quad \text{and} \quad r_{YZ} = 0.10,$$

thus:

$$r_{XZ} = (0.05 + 0.10)/(1 + 4 \times 0.05 \times 0.10) = 0.147,$$

and translating into distances:

$$d_{XY} = 5.017\,\text{cM} \quad \text{and} \quad d_{YZ} = 10.137\,\text{cM},$$

thus:

$$d_{XZ} = 5.017 + 10.137 = 15.154\,\text{cM}.$$

The value 15.154 cM translates back to $r_{XZ} = 0.147$:

$$r_{XZ} = 0.5 \tanh(0.02 \times 15.154) = 0.147,$$

the same value as the direct calculation.

Many more mapping functions have been developed that fitted better to experimental data or for specific situations encountered in genetic research. For common situations, however, the Kosambi and Haldane mapping functions are considered adequate. Figure 5.1 illustrates the relationship between map distance and

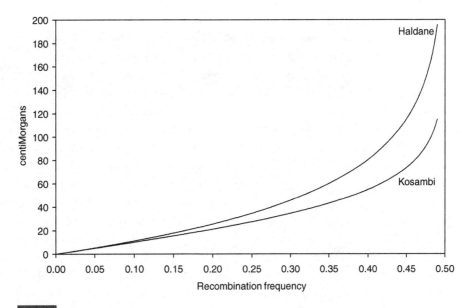

Fig. 5.1

Relationship between recombination frequency and map distance for the Haldane and the Kosambi mapping functions.

recombination frequency under both mapping functions. It shows that the difference between the two functions is very small for distances under 10 cM. Thus, when studying local map regions, the choice of map function is not very relevant. However, when studying longer regions or total map length, the choice becomes important. For practical applications of linkage maps such as QTL analysis, marker-assisted breeding and map-based cloning, which concentrate on shorter map regions, the choice between the Kosambi and the Haldane mapping function is not a real issue. Finally, Haldane's mapping function is the only mapping function that allows easy calculation of so-called multipoint likelihoods (Thompson, 2000), which will be addressed in Section 5.4.

5.2 An example experiment

Before we continue with the estimation of a map, we will describe a simple experiment that will become the subject of the mapping procedures. Consider a backcross population segregating for three successive loci, **A**, **B** and **C**, as the example experiment:

$(F_1) \, ABC/abc \times (P_1) \, abc/abc$.

Table 5.1. Simulated backcross for 100 individuals with three segregating loci.

	Genotype			No. of	Recombination in segment		
Code	A	B	C	individuals	A–B	B–C	A–C
ABC	Aa	Bb	Cc	43	No	No	No
ABc	Aa	Bb	cc	4	No	Yes	Yes
AbC	Aa	bb	Cc	0	Yes	Yes	No
Abc	Aa	bb	cc	4	Yes	No	Yes
aBC	aa	Bb	Cc	3	Yes	No	Yes
aBc	aa	Bb	cc	1	Yes	Yes	No
abC	aa	bb	Cc	1	No	Yes	Yes
abc	aa	bb	cc	44	No	No	No
Total				100			

In a backcross with three segregating loci, there are 2^3, or eight, possible genotypes. This backcross was simulated in a computer according to Mendelian inheritance with independence of recombination in adjacent segments (thus (5.1) applies), whilst the locus order was **A–B–C**. Locus genotypes were obtained for 100 individuals; a summary is given Table 5.1. Application of the theory of Chapter 3 provides us with the following direct pairwise estimates of the three recombination frequencies:

$$r_{AB} = 0.08, \; r_{BC} = 0.06 \quad \text{and} \quad r_{AC} = 0.12.$$

The sum of adjacent recombination frequencies is $0.08 + 0.06 = 0.14$, which, according to the theory, is indeed greater than the recombination frequency between **A** and **C**. We know that we should correct this value with twice the product of the adjacent recombination frequencies (5.1). This results in:

$$0.08 + 0.06 - 2 \times 0.08 \times 0.06 = 0.1304,$$

which is different from the direct estimate, $r_{AC} = 0.12$. What are the reasons for this? First, the experiment represents a finite sample. The 100 individuals in the population are the result of recombination events sampled according to Mendelian rules and predefined map distances. This means that the data are subject to sampling variation. Secondly, recombination is a discrete event: either there is recombination or there is not. This is in contrast to (5.1), which predicts the recombination frequency between **A** and **C** based on the two neighbouring segment frequencies, but it is of course impossible in our example to obtain a set of 13.04 recombinant individuals. Similarly, double recombination events are predicted to occur with a probability of 0.0048 (meaning 0.48 individual), whilst actually one was observed (at genotype *aa,Bb,cc*). Increasing the size of the population will reduce the discrepancies that are due to sampling variation and to recombinations being discrete events. These data were simulated under a model where recombination events in neighbouring segments are mutually independent. If, instead of independence, a certain level of crossover interference had been present in the recombination, then the above discrepancies would be expected to be larger, because equation (5.1) would not apply.

5.3   Map estimation by linear regression

When we perform a mapping experiment, we observe locus genotypes on all individuals. From these data, we may estimate pairwise recombination frequencies, from which we want to estimate a map in additive centiMorgan units. There are two major approaches to map estimation. The first is to translate recombination frequencies into map distances using a mapping function and subsequently work with a linear model. This approach was introduced by Jensen & Jørgenson (1975) and later (independently) by Lalouel (1977). The second is to set up a likelihood model for the locus observations on all individuals and estimate the recombination frequencies by maximization of the likelihood, and by subsequently translating recombination frequencies into map distances. In this section, map estimation using the linear model according to Jensen & Jørgenson (1975) is described.

The advantage of the property of additivity of map distances is that it is straightforward to set up a linear model. Suppose we are dealing with three loci in the order **A–B–C**, then we are able to obtain three pairwise distance estimates, d_{AB}, d_{BC} and d_{AC}. However, the map is determined by only the two adjacent distances: δ_{AB} and δ_{BC}. We may define the following model:

$$
\begin{aligned}
d_{AB} &= & \delta_{AB} + e_{AB} &= 1 \times \delta_{AB} + 0 \times \delta_{BC} + e_{AB} \\
d_{BC} &= & \delta_{BC} + e_{BC} &= 0 \times \delta_{AB} + 1 \times \delta_{BC} + e_{BC} \\
d_{AC} &= \delta_{AB} + \delta_{BC} + e_{AC} &= 1 \times \delta_{AB} + 1 \times \delta_{BC} + e_{AC}
\end{aligned}
$$

or, in matrix notation:
$$
\begin{pmatrix} d_{AB} \\ d_{BC} \\ d_{AC} \end{pmatrix} = \begin{pmatrix} 1 & 0 \\ 0 & 1 \\ 1 & 1 \end{pmatrix} \begin{pmatrix} \delta_{AB} \\ \delta_{BC} \end{pmatrix} + \begin{pmatrix} e_{AB} \\ e_{BC} \\ e_{AC} \end{pmatrix},
$$

in which e_{AB}, e_{BC} and e_{AC} are deviations. This is a standard linear model with three equations and two unknown parameters. The system can be solved by the *least squares* method; the method minimizes the sum of the squared deviations and is usually called *linear regression*. The model can easily be extended to incorporate more loci. (*Remark*: it can also be used to combine results from different populations, which may be used for map integration.)

In general, a distance is modelled as the sum of all constituent adjacent distances plus an error term. For instance, suppose the order is **B–A–C**, then the map will be determined by the two constituent distances, δ_{BA} and δ_{AC}. Because there is no direction in a distance, the locus indices may be exchanged; thus: $\delta_{BA} = \delta_{AB}$. The linear model corresponding to order **B–A–C** reads:

$$
\begin{aligned}
d_{AB} = & \quad \delta_{AB} + e_{AB} = 1 \times \delta_{AB} + 0 \times \delta_{AC} + e_{AB} \\
d_{BC} = & \; \delta_{AB} + \delta_{AC} + e_{BC} = 1 \times \delta_{AB} + 1 \times \delta_{AC} + e_{BC} \\
d_{AC} = & \quad \delta_{AC} + e_{AC} = 0 \times \delta_{AB} + 1 \times \delta_{AC} + e_{AC}
\end{aligned}
$$

or, in matrix notation:
$$
\begin{pmatrix} d_{AB} \\ d_{BC} \\ d_{AC} \end{pmatrix} = \begin{pmatrix} 1 & 0 \\ 1 & 1 \\ 0 & 1 \end{pmatrix} \begin{pmatrix} \delta_{AB} \\ \delta_{AC} \end{pmatrix} + \begin{pmatrix} e_{AB} \\ e_{BC} \\ e_{AC} \end{pmatrix}.
$$

As an illustration, we can apply this model to the data of the experiment described in the previous section, for map order **A–B–C**. First, we have to obtain the map distances by applying the Haldane mapping function to the estimated recombination frequencies:

$$
d_{AB} = -50 \ln(1 - 2 \times 0.08) = 8.72 \text{ cM}, \quad d_{BC} = 6.39 \quad \text{and} \quad d_{AC} = 13.72.
$$

The model thus becomes:

$$
\begin{pmatrix} 8.72 \\ 6.39 \\ 13.72 \end{pmatrix} = \begin{pmatrix} 1 & 0 \\ 0 & 1 \\ 1 & 1 \end{pmatrix} \begin{pmatrix} \delta_{AB} \\ \delta_{BC} \end{pmatrix} + \begin{pmatrix} e_{AB} \\ e_{BC} \\ e_{AC} \end{pmatrix}.
$$

Solving this with the method of least squares results in the following map:

$$
\hat{\delta}_{AB} = 8.26 \quad \text{and} \quad \hat{\delta}_{BC} = 5.93.
$$

One of the standard assumptions in linear regression is that the error terms have equal variance. This, however, is not the case here: large distances are estimated with much lower precision than small distances. This fact can easily be seen from the estimators of the corresponding recombination frequencies in a backcross: as they are binomial variables, they have a variance equal to $r\,(1 - r)/n$, with n as the number of individuals. In order to adjust for unequal variances, the model can be solved with *weighted least squares* using the inverted variance estimates of the distances as weights. Because the model is in *distances* rather than recombination frequencies, Jensen & Jørgenson (1975) and Lalouel (1977) used an approximating formula to get the distance variances from the recombination frequency variances. When the recombination frequency is estimated as zero, which is quite common in practical situations, the corresponding variance will also be estimated as zero. Unfortunately, these zero values cannot be used as weights in the weighted least squares. In the regression-mapping algorithm as implemented in JoinMap, this is solved heuristically by taking the square of the *LOD*s as weights (Van Ooijen, 2006). (*Remark*: in the original JoinMap algorithm, the *LOD* value itself was used instead of its square (Stam, 1993); later experience showed that better results were obtained with the square of the *LOD*s (Stam, personal communication).) Applying this latter method to our example, for which the *LOD*s are:

$$LOD_{AB} = 18.00 \,, \quad LOD_{BC} = 20.25 \quad \text{and} \quad LOD_{AC} = 14.17 \,,$$

a slightly different map is estimated:

$$\hat{\delta}_{AB} = 8.31 \quad \text{and} \quad \hat{\delta}_{BC} = 6.07 \text{ (vs 8.26 and 5.93)} \,.$$

Another assumption in linear regression is that the error terms are mutually independent. If the distances are obtained from different segregating populations, then this is a valid assumption. Most of the pairwise data analysed by Jensen & Jørgenson (1975) came from independent experiments, and when this was not the case, they made sure to use data only from non-overlapping map segments, of which the dependence may be neglected. However, if the distances are obtained from the same population, then the overlapping distance estimates are correlated. A proper way to deal with the situation is to solve the linear model with *generalized least squares*, in which the covariance structure of all the distance estimates is used. However, the covariance structure depends on the map order, so when many map orders are to be evaluated (as we shall see later), then adjusting the covariance structure for each map order will make the computations very time consuming. Lalouel (1977) ignores the independence assumption, even though it appears to be violated occasionally. Weeks & Lange (1987), in their investigation of the possibilities of using Lalouel's method as a mapping procedure preliminary to the maximum likelihood mapping, also disregard the assumption. The regression-mapping algorithm as implemented in JoinMap also ignores the dependence of distance estimates. In computer simulations, this appeared to have no big effect (Stam, 1993 and personal communication).

Missing genotype observations

What happens in this algorithm if there are missing genotype observations? If the individuals with missing observations do not have recombination events in the map segments involved, then there will be no disturbing effect. However, with denser maps, this is not very likely to occur. If in a segment **A–B–C** an individual has recombination between **A** and **B** and the genotype of **B** is not observed, then the estimate of d_{AB} will be reduced whereas d_{AC} remains unchanged. This will result in a poorer fit of the regression model. In general, a higher number of missing observations will lead to a poorer fit.

Crossover interference

What happens if there is a certain level of crossover interference? The regression model is based on distances obtained from estimates of pairwise recombination frequencies translated with a mapping function. It is straightforward to use the Kosambi instead of the Haldane mapping function. For instance, with our backcross example, the pairwise map distances become:

$$d_{AB} = 25 \ln((1 + 2 \times 0.08)/(1 - 2 \times 0.08)) = 8.07\,\text{cM},$$
$$d_{BC} = 6.03\,\text{cM} \quad \text{and} \quad d_{AC} = 12.24\,\text{cM}.$$

With the linear regression model for the order **A–B–C**, we obtain the following map:

$$\hat{\delta}_{AB} = 7.45 \quad \text{and} \quad \hat{\delta}_{BC} = 5.41.$$

An interesting aspect here is that applying a certain mapping function in fact models the level of crossover interference, enabling the prediction of the recombination frequency of the overall segment from the recombination frequencies of the constituent segments. Thus, if the right mapping function is applied in the regression model, then it should show the best fit, for instance the smallest residual sum of squares. The data of the example were generated with independence of recombination in adjacent segments, i.e. based on the applicability of the Haldane mapping function. If we compare the residual sums of squares of the backcross example for the order **A–B–C**, we will find that this is 0.64 for the Haldane mapping function and 1.15 for the Kosambi mapping function, which agrees with the way the data were simulated.

5.4 Map estimation by maximum likelihood

The first application of maximum likelihood in linkage studies is due to Haldane & Smith (1947), who considered the estimation of recombination frequencies for pairs of loci in human pedigrees. In genetics, the application of maximum likelihood involving all available data of a general pedigree or a set of more than two loci is usually referred to as multipoint maximum likelihood.

By definition, the likelihood is the joint probability of the observations given a model; the likelihood is considered as a function of the model parameters. In our case of map estimation, the model is defined by the order of the loci and the recombination frequencies between adjacent loci. Because the individuals are independent, the joint probability is the product of the probabilities of individuals. Formally, this should be multiplied by the multinomial coefficient, as the set of individuals is unordered. Being a constant factor, the multinomial coefficient does not affect the estimation and therefore it is usually omitted. The probability of a particular individual depends on its phenotype. Let us continue with the example of three loci, **A**, **B** and **C**, with map order **A–B–C** in a backcross population. The recombination frequencies relevant for the model are the ones between **A** and **B**, r_{AB}, and between **B** and **C**, r_{BC}. We will use the genotype codes of Table 5.1. Any individual with genotype ABC is the result of having no recombination in segment **A–B** as well as no recombination in segment **B–C**. The probability of genotype Aa is $\frac{1}{2}$, the probability of no recombination in **A–B** is $(1 - r_{AB})$ and the probability of no recombination in **B–C** is $(1 - r_{BC})$. Thus, assuming independence of recombination events in adjacent intervals, the probability of genotype ABC is:

$$P_{ABC} = \tfrac{1}{2}(1 - r_{AB})(1 - r_{BC}).$$

The probability for an individual with genotype ABc is:

$$P_{ABc} = \tfrac{1}{2}(1 - r_{AB})r_{BC}.$$

This is the result of no recombination in segment **A–B** (with probability $(1 - r_{AB})$) and recombination in segment **B–C** (with probability r_{BC}). Probabilities for the other genotypes can be obtained in the same way. If we denote the number of observations of a given genotype g by N_g and the total number of individuals by N_t, then the likelihood is:

$$L = \prod_g P_g{}^{N_g},$$

and the log-likelihood is:

$$\ln(L) = \sum_g N_g \ln(P_g).$$

The log-likelihood can be written as:

$$\ln(L) = (N_{ABC} + N_{ABc} + N_{abC} + N_{abc})\ln(1 - r_{AB}) +$$
$$(N_{AbC} + N_{Abc} + N_{aBC} + N_{aBc})\ln(r_{AB}) +$$
$$(N_{ABC} + N_{aBC} + N_{Abc} + N_{abc})\ln(1 - r_{BC}) +$$
$$(N_{ABc} + N_{aBc} + N_{AbC} + N_{abC})\ln(r_{BC}) - N_t \ln(2).$$

If we inspect this formula and look at the coefficient before the term $\ln(1 - r_{AB})$ in this equation, we can see that this is the total number of genotypes *without* recombination in segment **A–B**. The coefficient before the term $\ln(r_{AB})$ is the total number of genotypes *with* recombination in segment **A–B**. The next two terms show similar forms: here, the coefficients are the total genotype numbers with respect to recombination in segment **B–C**. We can define R_{AB} as the number of individuals with recombination in segment **A–B**, and S_{AB} as $N_t - R_{AB}$, and similarly define R_{BC} and S_{BC} for segment **B–C**. If we also define s_{AB} as $1 - r_{AB}$ and s_{BC} as $1 - r_{BC}$, then the log-likelihood becomes:

$$\ln(L) = S_{AB}\ln(s_{AB}) + R_{AB}\ln(r_{AB}) + S_{BC}\ln(s_{BC}) + R_{BC}\ln(r_{BC}) - N_t \ln(2).$$

In order to estimate the map, we must find the values of r_{AB} and r_{BC} that maximize the log-likelihood (and thus the likelihood). Because the partial derivative of the log-likelihood with respect to r_{AB} does not depend on r_{BC} and vice versa, the solutions are identical to the pairwise or two-point estimators of the recombination frequencies obtained in Chapter 3:

$$\hat{r}_{AB} = R_{AB}/N_t \quad \text{and} \quad \hat{r}_{BC} = R_{BC}/N_t.$$

To finalize the map estimation procedure, all we need to do is translate these two recombination frequencies into map distances. If we apply this to our backcross example, we obtain:

$$\hat{r}_{AB} = 8/100 = 0.08 \quad \text{and} \quad \hat{r}_{BC} = 6/100 = 0.06,$$

and

$$\ln(L) = 92\ln(0.92) + 8\ln(0.08) + 94\ln(0.94) + 6\ln(0.06) - 100\ln(2)$$
$$= -119.89.$$

Using the Haldane mapping function, the estimated map distances are:

$$\hat{\delta}_{AB} = 8.72 \quad \text{and} \quad \hat{\delta}_{BC} = 6.39.$$

If we are supposing another map order, then the model changes. For instance, if we suppose the order **B–A–C**, then the likelihood will be determined by the two recombination frequencies $r_{BA} = r_{AB}$ and r_{AC}. The corresponding log-likelihood can be derived as:

$$\ln(L) = S_{AB}\ln(s_{AB}) + R_{AB}\ln(r_{AB}) + S_{AC}\ln(s_{AC}) + R_{AC}\ln(r_{AC}) - N_t\ln(2)$$

and the recombination frequency estimators are:

$$\hat{r}_{AB} = R_{AB}/N_t \quad \text{and} \quad \hat{r}_{AC} = R_{AC}/N_t \,.$$

Missing genotype observations

In the above situations, all genotype information is present. In these cases, the multipoint maximum likelihood estimates of the recombination frequencies are identical to the two-point estimates. How will missing observations affect multipoint maximum likelihood estimation? The major consequence is that the likelihood becomes more complicated. We return to our backcross example and suppose that the map order is **A–B–C**. Suppose that for an individual the phenotype AuC is observed, i.e. the score at locus **B** is missing (u stands for unknown). This means that the phenotype has two possibilities for the underlying genotype: ABC and AbC. Then the probability for the phenotype becomes the sum of the two genotype probabilities:

$$P_{AuC} = P_{ABC} + P_{AbC} = \tfrac{1}{2}(1 - r_{AB})(1 - r_{BC}) + \tfrac{1}{2}r_{AB}\,r_{BC}\,.$$

The contribution of the individual to the log-likelihood becomes:

$$\ln(P_{AuC}) = \ln((1 - r_{AB})(1 - r_{BC}) + r_{AB}\,r_{BC}) - \ln(2)\,.$$

In order to maximize the log-likelihood, we must take the partial derivatives with respect to the parameters r_{AB} and r_{BC}. It is not difficult to see that now the partial derivative with respect to r_{AB} depends on r_{BC} and vice versa. The effect of this is that explicit estimators for the unknown recombination frequencies can no longer be obtained. Estimates can be obtained using the EM algorithm, which can be implemented analytically as in MAPMAKER (Lander & Green, 1987; Lander *et al.*, 1987) or stochastically using *Monte Carlo* methods (i.e. *Gibbs sampling*) as in JoinMap (Van Ooijen, 2006). An important result for when there are missing observations is that multipoint estimates of recombination frequencies will be different from pairwise or two-point estimates. Multipoint estimates are considered more accurate, because the information from all loci is taken into account.

Crossover interference

The multipoint maximum likelihood approach assumes independence of recombination in adjacent segments, i.e. absence of crossover interference. If we are strict, map distances should be obtained using the Haldane mapping function. However, the software MAPMAKER, which is based on multipoint maximum likelihood, also allows the use of the Kosambi mapping function (Lander *et al.*, 1987). Setting up a simple likelihood model incorporating crossover interference is not feasible (Thompson, 2000). Speed *et al.* (1992) investigated the robustness of the above likelihood model for ordering loci in the presence of crossover interference. They provided proof that, even when there is crossover interference, the order that maximizes the no-interference likelihood is a consistent estimator of the true order. This is reassuring with respect to the order, but it is difficult to say what effect crossover interference has on the estimation of map distances.

Exercises

Exercise 5.1. Estimating a map with linear regression
For three loci, **A**, **B** and **C**, three pairwise recombination frequencies are estimated: 0.08, 0.06 and 0.12 between **A** and **B**, **B** and **C**, and **A** and **C**, respectively. With MS Excel, it is not possible to do weighted linear regression, but you can do standard linear regression with the internal function LINEST(). Estimate the maps for the three possible orders using LINEST(). (*Remark*: *save the spreadsheet file for the next exercise.*)

Exercise 5.2. Estimating a map with linear regression
For four loci, **A**, **B**, **C** and **D**, six pairwise recombination frequencies are estimated:

$$\hat{r}_{AB} = 0.140, \hat{r}_{AC} = 0.257, \hat{r}_{AD} = 0.187,$$
$$\hat{r}_{BC} = 0.179, \hat{r}_{BD} = 0.069, \hat{r}_{CD} = 0.104.$$

Let us assume that **A** and **C** are the loci on the outside. Estimate the maps for the two orders **A–B–D–C** and **A–D–B–C** using LINEST(). (*Remark: the four markers are actually the same loci from Exercise 3.2. Save the spreadsheet file for the next exercise.*)

Exercise 5.3. Estimating a map with weighted least squares

This exercise should be done in the R statistical package. We will explain the necessary commands in R using the linear regression as described in Section 5.3 for loci **A**, **B** and **C** of the example experiment. In R, after the command prompt '>', you can define the Haldane mapping function with the following code (the spaces are optional and will not affect the program):

```
> haldane = function(rf) {-50.0*log(1 - 2*rf)}
```

The vector with three pairwise recombination frequencies, RFs, can be created and assigned values as follows:

```
> RFs <- c(0.08, 0.06, 0.12)
```

Typing just a variable name, like RFs, as a command will make R show the contents of the variable (note that R is case sensitive, e.g. the variable 'rfs' will be different from 'RFs'; the line starting [1] indicates output):

```
> RFs
[1] 0.08 0.06 0.12
```

Applying the defined haldane() function to this vector produces the vector with pairwise distances, PDs, in centiMorgan units:

```
> PDs <- haldane(RFs)
```

Next, we create the two design vectors X1 and X2 that apply to the order **A–B–C** as in Section 5.3, i.e. the first and second column of the matrix shown there:

```
> X1 <- c(1, 0, 1)
> X2 <- c(0, 1, 1)
```

The linear model can be defined, solved and stored in the variable LS with the lm() function:

```
> LS <- lm(PDs ~ X1 + X2 - 1)
```

The '−1' in the model definition 'PDs ~ X1 + X2 − 1' means that no intercept should be fitted, which is what we want. The results can be shown with the summary() function and the analysis of variance table with the anova() function:

```
> summary(LS)
[output]
> anova(LS)
[output]
```

If we wish to apply *weighted* least squares with the squared *LOD*s as weights, we can create the vector with the *LOD*s. From this vector, calculate the squared LODs as the weights vector LOD2 and use this in the weights-option of the lm() function:

```
> LOD <- c(18.00, 20.25, 14.17)
> LOD2 <- LOD^2
> WLS <- lm(PDs ~ X1 + X2 - 1, weights = LOD2)
```

This will produce the results as presented in Section 5.3.

The *LOD*s for the pairs of the four loci **A**, **B**, **C** and **D** of the previous exercise are:

$$LOD_{AB} = 17.05, \quad LOD_{AC} = 7.26, \quad LOD_{AD} = 12.75,$$
$$LOD_{BC} = 14.03, \quad LOD_{BD} = 27.85, \quad LOD_{CD} = 22.45.$$

Estimate the maps for the two orders **A–B–D–C** and **A–D–B–C** with weighted least squares using the squared *LOD*s as weights.

Exercise 5.4. Estimating a map with maximum likelihood

In a normal backcross population (F_1) *DEF / def* × (P_1) *def / def*, the genotypes of loci **D**, **E** and **F** were observed for 100 plants and the following combined genotypes were counted:

Code	Genotype			No. of individuals
	D	E	F	
DEF	Dd	Ee	Ff	40
DEf	Dd	Ee	ff	0
DeF	Dd	ee	Ff	7
Def	Dd	ee	ff	5
dEF	dd	Ee	Ff	9
dEf	dd	Ee	ff	8
deF	dd	ee	Ff	2
def	dd	ee	ff	29
Total				100

Calculate the log-likelihood for the map order **D–E–F**. (*Remark: save the spreadsheet file for the next exercise.*)

References

Haldane, J. B. S. (1919). The combination of linkage values, and the calculation of distance between the loci of linked factors. *Journal of Genetics*, 8, 299–309. http://www.ias.ac.in/jarch/jgenet/8/JG_8_299.pdf.

Haldane, J. B. S. & Smith, C. A. B. (1947). A new estimate for the linkage between the genes for colour-blindness and haemophilia in man. *Annals of Eugenics*, 14, 10–31. http://onlinelibrary.wiley.com/doi/10.1111/j.1469-1809.1947.tb02374.x/pdf.

Jensen, J. & Jørgensen, J. H. (1975). The barley chromosome 5 linkage map. I. Literature survey and map estimation procedure. *Hereditas*, 80, 5–16.

Kosambi, D. D. (1944). The estimation of map distances from recombination values. *Annals of Eugenics*, 12, 172–5. http://kkvinod.webs.com/share/Kosambi_1944.pdf.

Lalouel, J. M. (1977). Linkage mapping from pair-wise recombination data. *Heredity*, 38, 61–77.

Lander, E. S. & Green, P. (1987). Construction of multilocus genetic linkage maps in humans. *Proceedings of the National Academy of Sciences USA*, 84, 2363–7.

Lander, E. S., Green, P., Abrahamson, J. *et al.* (1987). MAPMAKER: an interactive computer package for constructing primary genetic linkage maps of experimental and natural populations. *Genomics*, 1, 174–81.

Speed, T. P., McPeek, M. S. & Evans, S. N. (1992). Robustness of the no-interference model for ordering genetic markers. *Proceedings of the National Academy of Sciences USA*, 89, 3103–6.

Stam, P. (1993). Construction of integrated genetic linkage maps by means of a new computer package: JoinMap. *The Plant Journal*, 3, 739–44.

Sturtevant, A. H. (1913). The linear arrangement of six sex-linked factors in *Drosophila*, as shown by their mode of association. *Journal of Experimental Zoology*, 14, 43–59. http://www.esp.org/foundations/genetics/classical/holdings/s/ahs-13.pdf.

Thompson, E. (2000). *Statistical Inference from Genetic Data on Pedigrees*. NSF-CBMS Regional Conference Series in Probability and Statistics, Vol. 6. Beachwood, OH: Institute of Mathematical Statistics.

Van Ooijen, J. W. (2006). *JoinMap® 4, Software for the Calculation of Genetic Linkage Maps in Experimental Populations*. Wageningen, The Netherlands: Kyazma B.V..

Weeks, D. E. & Lange, K. (1987). Preliminary ranking procedures for multilocus ordering. *Genomics*, 1, 236–42.

6 Criteria for the evaluation of maps

The preceding chapter explained how to estimate a map for a given order. The problem is that we do not know which order is the correct one. The order of the loci on the map cannot be observed directly but must be inferred from the phenotype observations. For any given order of the loci, we can estimate a map and subsequently calculate the value of a criterion measuring the quality of the result. The best order can then be chosen based on the criterion. This chapter investigates several criteria suitable for this purpose.

6.1 Introduction

In the previous chapter, two methods were described for the estimation of a map for a given order. However, the order of loci cannot be observed directly. From the genotype observations of the loci, the pairwise (i.e. two-point) recombination frequencies can be estimated, and using these estimates, the map order must be inferred. The first type of experiment that was used for this purpose was the *three-point testcross*: an individual heterozygous at three loci is crossed with a homozygous *tester*. It is equivalent to a backcross segregating for three loci, for instance **A**, **B** and **C**:

$$(\text{F}_1) \, ABC/abc \times (\text{P}_1) \, abc/abc.$$

We will use an example, the genotypes of which have been summarized in Table 6.1. Using the theory of Chapter 3, we estimate the pairwise recombination frequencies by:

$$\hat{r}_{AB} = 0.16, \quad \hat{r}_{AC} = 0.08 \quad \text{and} \quad \hat{r}_{BC} = 0.20.$$

Because the recombination frequency between **B** and **C** is the largest of the three estimates, we conclude that these two loci are furthest apart, and therefore the most likely order is **B–A–C** (or, equivalently **C–A–B**). For a three-point testcross, the criterion that we can use to determine the best order is simply the largest pairwise recombination frequency. However, if more than three loci segregate in this way,

Table 6.1. Example results of a three-point testcross on 100 individuals.

Genotype				No. of individuals	Recombination in segment		
Code	A	B	C		A–B	B–C	A–C
ABC	*Aa*	*Bb*	*Cc*	42	No	No	No
ABc	*Aa*	*Bb*	*cc*	3	No	Yes	Yes
AbC	*Aa*	*bb*	*Cc*	8	Yes	Yes	No
Abc	*Aa*	*bb*	*cc*	0	Yes	No	Yes
aBC	*aa*	*Bb*	*Cc*	2	Yes	No	Yes
aBc	*aa*	*Bb*	*cc*	6	Yes	Yes	No
abC	*aa*	*bb*	*Cc*	3	No	Yes	Yes
abc	*aa*	*bb*	*cc*	36	No	No	No
Total				100			

this simple criterion cannot be applied to four or more loci. Of course, the criterion may be helpful in finding the best overall order by considering all subsets of three loci. However, the criterion may be inconclusive with regard to finding the best order for all loci. In the past, several criteria have been proposed and are currently used in practical applications. The most common criteria will be described in this chapter. The criteria will be demonstrated using the three-point testcross example of Table 6.1.

6.2 Log-likelihood

When estimating the map with the maximum likelihood method, the obtained likelihood, or equivalently the log-likelihood, can be used to compare different orders. According to Chapter 5, the value of the log-likelihood for map order **A–B–C** in the three-point testcross is equal to:

$$\ln(L_{A-B-C}) = S_{AB} \ln(s_{AB}) + R_{AB} \ln(r_{AB}) + \\ S_{BC} \ln(s_{BC}) + R_{BC} \ln(r_{BC}) - N_t \ln(2),$$

in which R_{XY} and S_{XY} represent the numbers of recombinant and non-recombinant individuals between loci **X** and **Y**, r_{XY} represents the maximum likelihood estimate of the recombination frequency between loci **X** and **Y**, $s_{XY} = 1 - r_{XY}$, and N_t represents the total number of individuals. For map order **A–B–C**, we obtain with the data from the example:

$$\ln(L_{A-B-C}) = 84\ln(0.84) + 16\ln(0.16) +$$
$$80\ln(0.80) + 20\ln(0.20) - 100\ln(2) = -163.3 .$$

For the two other orders of the example, we obtain:

$$\ln(L_{A-C-B}) = -147.2 \quad \text{and} \quad \ln(L_{B-A-C}) = -141.2 .$$

Based on the value of the log-likelihood criterion, locus order **B–A–C** provides the best map, i.e. the map with the least negative log-likelihood value. This is in agreement with the result of the maximum recombination frequency criterion for the three-point testcross. In this example with complete segregation information, calculation of the log-likelihood value is straightforward. In other situations, multipoint estimates have to be obtained and used in the computation of the log-likelihood, but this usually occurs simultaneously in the optimization process.

6.3 Sum of *LODs* of adjacent segments

Instead of using the natural logarithm of the likelihood, it is equivalent to use the common or base 10 logarithm. For map order **A–B–C**, the log-likelihood using the common logarithm reads:

$$\log_{10}(L_{A-B-C}) = S_{AB}\log_{10}(s_{AB}) + R_{AB}\log_{10}(r_{AB}) +$$
$$S_{BC}\log_{10}(s_{BC}) + R_{BC}\log_{10}(r_{BC}) - N_t\log_{10}(2) .$$

The common logarithm is used in the *LOD* score, which in the case of two loci, **X** and **Y**, is given by (cf. Section 4.3):

$$LOD_{XY} = S_{XY}\log_{10}(s_{XY}) + R_{XY}\log_{10}(r_{XY}) + N_t\log(2) .$$

We can rewrite the order **A–B–C** log-likelihood:

$$\log_{10}(L_{A-B-C}) = S_{AB}\log_{10}(s_{AB}) + R_{AB}\log_{10}(r_{AB}) + N_t\log_{10}(2) +$$
$$S_{BC}\log_{10}(s_{BC}) + R_{BC}\log_{10}(r_{BC}) + N_t\log_{10}(2) -$$
$$3 N_t\log_{10}(2) .$$

Thus:

$$\log_{10}(L_{A-B-C}) = LOD_{AB} + LOD_{BC} - 3 N_t\log_{10}(2) .$$

Thus, the base 10 log-likelihood is the sum of the *LODs* of adjacent segments minus a constant. Obviously, a different order will correspond to a different sum; for example, for the order **A–C–B**:

$$\log_{10}(L_{A-C-B}) = LOD_{AC} + LOD_{BC} - 3 N_t\log_{10}(2) .$$

Therefore, if we are looking for the order with the highest likelihood, we can also look for the order with the largest sum of *LOD*s of adjacent segments (termed *SALOD*). The *LOD*s for the pairs in the three-point testcross example are given by:

$$LOD_{AB} = 11.0, \quad LOD_{AC} = 18.0 \quad \text{and} \quad LOD_{BC} = 8.4.$$

For the *SALOD* for the three possible orders, we obtain:

$$SALOD_{A-B-C} = 11.0 + 8.4 = 19.4,$$
$$SALOD_{A-C-B} = 18.0 + 8.4 = 26.4 \quad \text{and}$$
$$SALOD_{B-A-C} = 11.0 + 18.0 = 29.0.$$

Based on the sum of *LOD*s of adjacent segments, the best order is the one with locus **A** in the middle.

In situations with incomplete information, the pairwise recombination frequency estimates deviate from their multipoint estimates. Because the pairwise estimates are the basis of the *LOD*s, there will be a bias in the results when used under such circumstances.

6.4   Sum of recombination frequencies in adjacent segments

The recombination frequency is essentially a probability. In Chapter 2, it was shown that the recombination frequency for loci that reside on the same chromosome is always less than $\frac{1}{2}$. This means that the probability of a recombination event is always smaller than the probability of not having a recombination event, and with a dense map it will be even smaller. Consequently, the order that requires the least number of recombination events has the highest probability. For instance, if a backcross individual has a recombination event between **A** and **C** and between **B** and **C**, whereas it has no recombination event between **A** and **B**, then its probability for the three possible orders will be:

$$P_{A-B-C} = \tfrac{1}{2} s_{AB} r_{BC},$$
$$P_{A-C-B} = \tfrac{1}{2} r_{AC} r_{BC}, \quad \text{and}$$
$$P_{B-A-C} = \tfrac{1}{2} s_{AB} r_{AC}.$$

Clearly, the order that requires two recombination events, **A–C–B**, has the smallest probability:

$$r_{AC} r_{BC} < s_{AB} r_{BC} \Leftrightarrow r_{AC} < s_{AB},$$

which is true because $r_{AC} < \frac{1}{2}$ and $\frac{1}{2} < s_{AB}$, and

$$r_{AC}\, r_{BC} < s_{AB}\, r_{AC} \;\Leftrightarrow\; r_{BC} < s_{AB}\,,$$

which is true because $r_{BC} < \frac{1}{2}$ and $\frac{1}{2} < s_{AB}$.

Note that the probabilities are determined by the recombination frequencies of only the adjacent segments for the given order. Similar relationships can be derived for other types of segregating populations. Of course, not just a single, but all individuals of the segregating population should be taken into account. Thus, any particular order that can be explained by the smallest number of recombination events over all individuals can be regarded as the most likely order. Counting the number of recombination events in adjacent segments for a certain map order over all individuals is equivalent to taking the sum of the recombination frequency estimates in adjacent segments (termed *SARF*).

If we apply the criterion to our three-point testcross example, we can use the pairwise recombination frequency estimates to calculate the sum for the three possible orders. We obtain:

$$SARF_{A-B-C} = 0.16 + 0.20 = 0.36\,,$$
$$SARF_{A-C-B} = 0.08 + 0.20 = 0.28 \quad \text{and}$$
$$SARF_{B-A-C} = 0.16 + 0.08 = 0.24\,.$$

Based on these sums, the best order is that with locus **A** in the middle, which is the same conclusion as before.

In many instances, the pairwise recombination frequency estimates will be different from the multipoint estimates, which are considered to be the more-accurate estimate (in the three-point testcross example, the two types of estimate are identical). Application of the *SARF* criterion based on different types of estimation may therefore result in different conclusions regarding the best order. In most situations, however, the differences between two-point and multipoint estimates are small. There is one major exception and that is when dealing with dominant markers in repulsion phase in the F_2. As we have seen in Chapter 3, such situations often result in two-point estimates of zero where the real value may be up to 0.2 or larger. In such instances, multipoint estimates should be used.

6.5   Residual sum of squares

If the map is estimated by (weighted) linear regression, then the solutions are obtained by the least squares method, which minimizes the residual sum of squares. If we wish to compare map orders (i.e. models), then we can use the residual sum of squares (SS_R) as the criterion. The smaller the residual sum of squares, the better the model fits the data. If we apply this to our three-point testcross example using the Haldane mapping function, we obtain the following:

$$\textbf{A–B–C: } \hat{\delta}_{AB} = 7.2\,\text{cM}, \ \hat{\delta}_{BC} = 13.5\,\text{cM}, \ SS_R = 435,$$
$$\textbf{A–C–B: } \hat{\delta}_{AC} = 3.7\,\text{cM}, \ \hat{\delta}_{BC} = 20.5\,\text{cM}, \ SS_R = 75 \quad \text{and}$$
$$\textbf{B–A–C: } \hat{\delta}_{AB} = 18.5\,\text{cM}, \ \hat{\delta}_{AC} = 7.9\,\text{cM}, \ SS_R = 2.$$

The smallest residual sum of squares is that of the map order **B–A–C**. If we use weighted linear regression, then the weighted residual sum of squares should be used. Applied to the example and using the square of the LODs as weights, we obtain the following:

$$\textbf{A–B–C: } \hat{\delta}_{AB} = 7.6\,\text{cM}, \ \hat{\delta}_{BC} = 5.5\,\text{cM}, \ SS_R = 51079,$$
$$\textbf{A–C–B: } \hat{\delta}_{AC} = 6.9\,\text{cM}, \ \hat{\delta}_{BC} = 17.2\,\text{cM}, \ SS_R = 8787 \quad \text{and}$$
$$\textbf{B–A–C: } \hat{\delta}_{AB} = 18.5\,\text{cM}, \ \hat{\delta}_{AC} = 8.4\,\text{cM}, \ SS_R = 237.$$

Also in this case, the order with locus **A** in the middle is the best. (Note that we also obtain different distance estimates than without weighing). As an alternative to the residual sum of squares, the multiple correlation coefficient of the (weighted) linear regression (usually indicated by R^2) may be used, with which the larger correlation is favoured.

6.6 χ^2 Goodness of fit

In the regression-mapping algorithm, it is possible to determine the fit of the pairwise recombination frequency estimates to the recombination frequencies that are obtained from the estimated map. This goodness-of-fit measure is a likelihood ratio statistic that compares all two-point recombination frequencies with the map-derived recombination frequencies. In addition to using it to compare maps, it also gives an impression of the overall quality of the map.

If we are dealing with a backcross, we know for every pair (p) of loci in the regression model the number of recombinant (R_p) and non-recombinant (S_p) gametes. Their numbers are equal to the product of the total number of gametes (N_p) and the two-point recombination frequency estimates r_{tp} and $s_{tp} = 1 - r_{tp}$, respectively. The value of N_p may vary across pairs due to differences in numbers of missing observations. From the estimated map, the recombination frequency of a pair of loci r_{mp} can be obtained by applying the inverse mapping function to the corresponding map distance. To compare the two-point recombination frequency estimates with the corresponding map-derived estimates over all pairs, we calculate the deviance as twice the natural logarithm of the likelihood ratio:

$$2\ln(LR) = 2\sum_p [\ln(L(r_{tp})/L(r_{mp}))]$$
$$= 2\sum_p [S_p \ln(s_{tp}) + R_p \ln(r_{tp}) - S_p \ln(s_{mp}) - R_p \ln(r_{mp})]$$
$$= 2\sum_p N_p[s_{tp} \ln(s_{tp}/s_{mp}) + r_{tp} \ln(r_{tp}/r_{mp})].$$

The goodness-of-fit measure, denoted as G^2, is distributed as a χ^2 with a number of degrees of freedom roughly equal to the difference between the number of pairs in the model (i.e. the number of equations) and the number of adjacent map distances (i.e. the number of unknown parameters). Because not all pairwise recombination frequency estimates are mutually independent, the statistic is only roughly distributed as a χ^2 variable. A poor goodness of fit corresponds to a large value of the test statistic. For other population types, the appropriate likelihood function has to be entered into the goodness-of-fit criterion. The criterion may also be used to investigate which mapping function leads to the best fit.

Applied to the three-point testcross example using the Haldane mapping function gives the following results:

A–B–C: $\hat\delta_{AB} = 7.2\,\text{cM}$, $\hat\delta_{BC} = 13.5\,\text{cM}$, $G^2 = 22$,
A–C–B: $\hat\delta_{AC} = 3.7\,\text{cM}$, $\hat\delta_{BC} = 20.5\,\text{cM}$, $G^2 = 5.6$ and
B–A–C: $\hat\delta_{AB} = 18.5\,\text{cM}$, $\hat\delta_{AC} = 7.9\,\text{cM}$, $G^2 = 0.1$.

In this case, there are three two-point estimates and two map distances; this means that the χ^2 distribution of the test statistic is based on one degree of freedom. The best goodness of fit is obtained for map order **B–A–C**. If weighted linear regression is used, the following results are obtained:

A–B–C: $\hat\delta_{AB} = 7.6\,\text{cM}$, $\hat\delta_{BC} = 5.5\,\text{cM}$, $G^2 = 37$,
A–C–B: $\hat\delta_{AC} = 6.9\,\text{cM}$, $\hat\delta_{BC} = 17.2\,\text{cM}$, $G^2 = 3.2$ and
B–A–C: $\hat\delta_{AB} = 18.5\,\text{cM}$, $\hat\delta_{AC} = 8.4\,\text{cM}$, $G^2 = 0.1$.

Again, the order with locus **A** in the middle is the best.

Exercises

Exercise 6.1.
This exercise continues with the three markers **A**, **B** and **C** of Exercise 5.1. The recombination frequencies were estimated from 100 backcross individuals.

A. What are the sums of recombination frequencies of adjacent segments for the three possible orders?
B. What are the residual sums of squares you obtained with LINEST()?
C. What are the χ^2 goodness-of-fit values?
D. Which is the best order?

Exercise 6.2.
This exercise continues with the four markers **A**, **B**, **C** and **D** of Exercise 5.2.

A. What are the sums of recombination frequencies of adjacent segments for the two orders **A–B–D–C** and **A–D–B–C**?
B. What are the residual sums of squares you obtained with LINEST()?
C. What are the χ^2 goodness-of-fit values?
D. Which is the best order?

Exercise 6.3.
This exercise continues with the three markers **D**, **E** and **F** of Exercise 5.4.

A. Estimate the three pairwise recombination frequencies.
B. There are three possible orders: **D–E–F**, **D–F–E** and **E–D–F**. For each order, calculate the pairwise recombination frequency between the outer-end markers by taking the sum of the comprising adjacent pairwise recombination frequencies, and compare these with the direct estimates of those recombination frequencies. What appears to be the best map order?
C. Do the same, but now taking the non-additivity of recombination frequencies into account. What appears to be the best map order?
D. Calculate the log-likelihoods for the three orders. What appears to be the best map order?
E. Calculate the number of recombination events for the three orders. What appears to be the best map order?
F. Calculate the *LOD* scores for the three pairs. Next, calculate for each order the sum of the *LOD*s of the adjacent segments. What appears to be the best map order?

Exercise 6.4.
We consider a backcross population ($XYZ / xyz \times xyz / xyz$) in which three markers, **X**, **Y** and **Z**, have been scored. The following data were obtained:

A. Calculate the recombination coefficients for the three marker pairs.
B. Which marker order should be preferred based on maximum likelihood?

	Phenotype			
Code	X	Y	Z	No. of individuals
XYZ	Xx	Yy	Zz	38
XYz	Xx	Yy	zz	0
XyZ	Xx	yy	Zz	7
Xyz	Xx	yy	zz	2
xYZ	xx	Yy	Zz	2
xYz	xx	Yy	zz	7
xyZ	xx	yy	Zz	4
xyz	xx	yy	zz	40
Total				100

C. Which marker order should be preferred based on the minimum number of recombinations between adjacent markers?

D. Which marker order should be preferred based on the maximum *SALOD*?

E. Calculate the χ^2 goodness-of-fit values for the three marker orders. Which marker order is the best?

7 How to find the best map order

As the number of loci increases, the number of possible locus orders soon becomes too large to calculate for each order the value of the evaluation criterion and then select the order with the best value. Smart ways of finding the best order are needed, capable of skipping many unnecessary evaluations. The problem of ordering loci is comparable to the well-known travelling salesman problem. Several algorithms that have been developed to solve this problem are also applied to the problem of ordering loci. This chapter gives a global description of these algorithms.

7.1 Introduction

In the previous two chapters, we learned how to estimate a map for a given order of loci and how to gauge the quality of the result. In order to find the best possible map, one is inclined to think of a simple and straightforward approach: calculate maps for all possible orders and simply compare them based on one of the evaluation criteria. This exhaustive approach is sometimes called a *brute force* search technique. For a few loci, this might be feasible, but if the number of loci increases, the number of possible orders soon becomes so large that even the fastest super-computer would not be able to do so within a lifetime.

Can we calculate the number of possible orders if we have l loci? Yes, at the first position of the map we may choose from l loci, at the second position the choice is reduced to $l - 1$ loci, at the third to $l - 2$, and so on until at the last position there is only one locus remaining. Because orders have no head or tail, any order is equivalent to its reverse. Consequently, for a given number of loci l the number of possible orders is:

$$\tfrac{1}{2} l(l - 1)(l - 2)\ldots 3 \times 2 \times 1 = \tfrac{1}{2} l!.$$

For a few typical values of l, Table 7.1 lists the number of possible locus orders. For ten loci, there are already 1.8 million possibilities. Currently, situations with more than 100 loci per linkage group are not uncommon. In such situations, the

Table 7.1. The number of possible map orders for l loci: $\frac{1}{2}\,l!$

l	$\frac{1}{2}\,l!$
3	3
5	60
10	1.8×10^6
20	1.2×10^{18}
50	1.5×10^{64}
100	4.7×10^{157}
1000	2.0×10^{2567}

number of possibilities becomes inconceivably large. To give an idea: 4.7×10^{157} is more than the number of grains of sand on the beach where you may go swimming. It is more than that of all the beaches in the world. It is more than all the atoms that the world consists of. It is more than all the atoms of our solar system, more than those of the Milky Way. In fact, more than the number of atoms of what it is estimated the entire universe consists of (which is in the order of 10^{80}; http://en.wikipedia.org/wiki/Observable_universe#Matter_content). Thus, the simple approach of comparing all possible orders is unfeasible, and we need smarter methods than a brute force search. Of course, common sense alone allows the exclusion of map orders in which closely linked loci (i.e. based on pairwise recombination frequency estimates) are placed far apart and, inversely, map orders in which loosely linked loci are at nearby positions. For maps with a larger number of loci, this may be insufficient to limit the huge number of possible orders. These situations require smart search algorithms that seriously reduce the many possibilities that must be evaluated.

7.2   The travelling salesman problem

The problem of finding the best map order is comparable to a well-known optimization problem, known as the travelling salesman problem. Imagine a salesman who has to visit clients in a number of cities to deliver his products. He leaves from home and wants to take the shortest possible route, visiting all clients before returning home (Fig. 7.1). In its simplest form, the distances between the cities define the problem. It may become more complicated, for instance, by taking into account extra costs (in time or money) involved in overtaking obstacles in certain

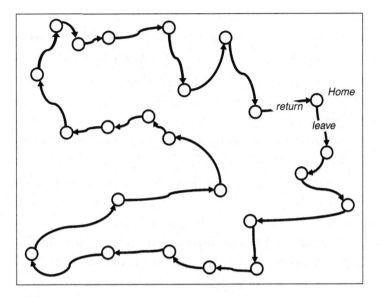

Fig. 7.1

The travelling salesman problem. A salesman has to visit his clients in various cities to deliver his products. He leaves from home and wants to take the shortest possible route to visit all clients and return home at the end. One of the possible routes is shown with arrows between the cities, shown as circles.

roads such as taking a ferry or crossing a mountain pass. But, whatever the complications, the best solution places the cities in a visiting order that presents the lowest overall costs (in distance, time or money).

The locus-ordering problem is similar. For a set of loci, we have their mutual distances; these may be recombination frequencies or genetic map distances. If we use the sum of adjacent distances as a map evaluation criterion, then we want the loci placed in the order with the smallest sum of adjacent distances (Fig. 7.2). However, there are also differences to the travelling salesman problem. The locus-ordering problem is one- rather than two-dimensional and the salesman has to return to where he started, whereas the genetic map has two different ends. The distances between the cities will be known quite precisely, whereas the recombination frequency estimates between loci will comprise a certain inaccuracy due to missing information (e.g. missing observations on locus genotypes, dominant genotype observations). The more-accurate multipoint recombination frequency estimates depend on the order, but the best order is not known beforehand, and thus for each order the distances may need to be re-estimated. Finally, recombination frequencies between loci are not linear but lie within the range from 0 to $\frac{1}{2}$;

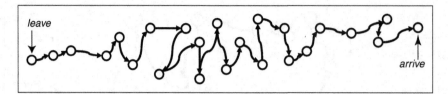

Fig. 7.2

The map order determination problem. The loci, shown as circles, are supposed to lie in one dimension, but because of inaccuracy in the distance estimates (i.e. recombination frequencies) are shown here to lie in two dimensions. We are looking for the shortest possible path through all loci, leaving at one end and arriving at the other. One of the possible routes is shown with arrows between the loci.

when translated to map distances, the recombination frequencies close to $\frac{1}{2}$ will go towards infinity, which forms an extra complication.

7.3 Optimization algorithms

Studying the travelling salesman problem belongs to a branch of mathematics called *combinatorial optimization* (Aarts & Lenstra, 2003; Applegate *et al.*, 2007). Various heuristic optimization methods can solve the travelling salesman problem (the Greek verb *heuriskein* means *to find* or *discover*; think of Archimedes' famous exclamation: Eureka!). Such procedures search the optimum based on common sense and experience but fail the mathematical proof of always finding the global optimum (Fig. 7.3). Slowly, new methods are seeping through to linkage analysis. Several have been applied to the locus-ordering problem and are often made available in computer software; some of this software is limited to the basic backcross population type. The methods include the following:

1. The *greedy* or *nearest-neighbour algorithm* (Buetow & Chakravarti, 1987; Stam, 1993; Ellis, 1999; Van Os *et al.*, 2005; Tan & Fu, 2006; Wu *et al.*, 2008).
2. *Simulated annealing* (Weeks & Lange, 1987; Jansen *et al.*, 2001; Jansen, 2005; Hackett *et al.*, 2003; Cartwright *et al.*, 2007).
3. *Tabu search* (Schiex & Gaspin, 1997).
4. The *genetic algorithm* (Gaspin & Schiex, 1997).
5. *Evolution strategy* (Mester *et al.*, 2003a, b).
6. *Ant colony optimization* (Iwata & Ninomiya, 2006).

We will describe the characteristics of the major methods below.

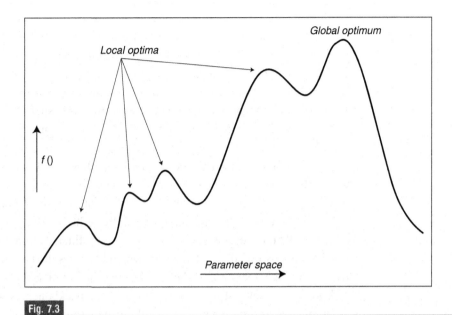

Fig. 7.3

Global optimum and some local optima in the to-be-optimized function $f()$.

7.4   Greedy algorithms

The greedy or nearest-neighbour algorithm is probably the most straightforward approach to finding the optimum map order. Basically, the greedy algorithm for the travelling salesman problem adds the city nearest to the current as the next city to travel to. The method is deterministic: repeated analyses of the same data always result in the same optimum. At present, possibly the most prominent version is that of Stam (1993). Stam's mapping algorithm builds the map locus by locus, starting from a pair of loci in the middle of the map. The next locus is chosen based on the sum of the pairwise LODs with all the loci that are already on the map. This criterion implies that the chosen locus has maximum linkage with the current loci. For this locus, the best position is determined given the positions of the current set of loci. Before continuing with the next locus, a local reshuffle of the order of the loci is carried out in order to prevent arriving at a local optimum. The method estimates map distances using weighted linear regression, using the square of the pairwise LODs as weights. A likelihood ratio goodness-of-fit measure not only determines the best position at which a locus is placed but is also used to verify whether adding the locus reduces the quality of the map so much that it is better not to add the locus to the map. The idea behind this is that any locus with erroneous observations will cause a poor fit.

Although the loci of one linkage group are supposed to lie in a one-dimensional space, i.e. the linkage map, missing and erroneous observations will cause the perceived system to be of a higher dimension (hence, the two dimensions of Fig. 7.2). The *minimum spanning tree* (MST) algorithm of Wu *et al.* (2008) starts with a greedy algorithm to compute the MST between all loci based on all pairwise recombination frequencies. This tree connects all loci in such a way that it has the shortest sum of lengths of connections, i.e. adjacent recombination frequencies. A tree, however, does not resemble a one-dimensional map, unless the tree has no branches. A branch in the tree means that the path at a certain locus on the main trunk splits off loci on the branch. Such a path cannot represent a one-dimensional genetic map. There is no guarantee, however, that the MST computations will result in a tree that is not branched, especially in higher-dimensional situations. Therefore, all branches will be resolved as follows. First, the longest (unbranched) path in the tree is found (i.e. the trunk), which becomes the backbone of the map. Next, one by one, loci on branches are inserted into the backbone at the location that leads to the smallest increase in length of the backbone. As this may lead to suboptimal map orders, the process is followed by three types of systematic order reshuffling until the map criterion (sum of adjacent recombination frequencies) stops improving. A subsequent step is a single EM iteration to impute missing observations and improve the recombination frequency estimates. The procedure then starts over again by computing the MST, and so on, until the EM step but now using the improved recombination frequencies. The procedure is repeated until convergence. It appears that the missing data imputation scheme leads to multipoint maximum likelihood estimates of the recombination frequency.

7.5   Simulated annealing

Kirkpatrick *et al.* (1983) established optimization by simulated annealing (SA). The procedure originates from physics and imitates the way atoms behave when a metal is cooled down. If the cooling is done gradually, then the material ends up as a strong, regular structure (i.e. the global optimum), whereas if the cooling is done fast, it becomes a brittle, irregular structure (i.e. a local optimum). Simulated annealing is a stochastic optimization technique: pseudo-random numbers determine the path through which the global optimum should be reached. A new map order is obtained by random mutation of the current order. Depending on the implementation of the algorithm, this might be a simple relocation of a single locus, or something more complex like an inversion of a range of loci. The new order is evaluated and subsequently accepted or rejected. Better orders are always accepted. In order to get out of local optima, poorer orders have a chance of becoming accepted. This chance depends on the magnitude of the decline of the optimization criterion, as well as on an *acceptance probability*. In the beginning, the acceptance probability is high: many poorer map orders will be accepted; this allows exploration of a large set of map orders. Gradually, the acceptance probability is decreased, until at the end only improvements are accepted. By analogy with the annealing metal, the acceptance probability is called the *temperature* that is gradually cooling down. The way the acceptance probability is reduced is called the *cooling schedule*.

A prominent implementation of SA is that of Jansen *et al.* (2001). The procedure minimizes the sum of recombination frequencies in adjacent segments. At first, these are two-point recombination frequencies, but once a map order is established, multipoint maximum likelihood recombination frequencies are obtained using Gibbs sampling. In the SA procedure, new map orders are obtained with simple steps of randomly sampling a locus that will be replaced to a random position. The acceptance probability is kept constant during a so-called *chain* of iterations (e.g. 1000), in which new map orders are generated by random locus replacements. In the subsequent chain, the acceptance probability is reduced by dividing it by $(1 + \alpha)$, with α as the so-called *cooling control* parameter with a small value, typically 0.001. The algorithm stops when a long series of chains has stopped delivering improvements. Because erroneous locus genotype observations may influence the algorithm towards poorer optima, an extra technique was introduced, called *spatial sampling*. With this method, loci are sampled that are spread over the entire map (based on the pairwise recombination frequencies only). This is founded on the notion that data errors have relatively less influence on larger distances. For the spatial sample of loci, a so-called *framework* map is calculated with the same SA procedure. In the next stage, the framework is used as a starting order to place the remaining loci.

7.6 Tabu search

A somewhat related method is tabu search. The idea behind *local search* techniques, such as SA, is that the global optimum can be reached by repeatedly making small changes from a given situation. Glover (1986) introduced the method to improve the efficiency of searching by including a mechanism that declares reversing recent changes as forbidden, i.e. they are *tabu*, in order to prevent returning to a previous situation. Gaspin & Schiex (1997) and Schiex & Gaspin (1997) have implemented tabu search for genetic mapping. They note that their algorithm needs further improvements, as it could not avoid becoming stuck at local optima. As the number of loci grows large, the number of possible changes will grow enormously, and presumably this may reduce the effectiveness of any tabu list of practical size.

7.7   Evolutionary algorithms

These algorithms are supposed to mimic how natural populations evolve under selection. Such procedures maintain a set, or population, of potential solutions instead of just a single candidate solution. This population will become improved in an iterative fashion by generating offspring solutions under mutation, recombination and selection. From the current set (of candidate solutions) of a given size, a new and larger set is created by making changes and/or by making combinations between the individual solutions. The new set is reduced to that of the original set by selecting the best candidates. If the new set is created by only making small random changes, then the method is called an *evolution strategy* (ES). Rechenberg (1973) introduced ES. If candidate solutions are also combined to create new solutions, then the method is called a *genetic algorithm* (GA). Holland (1975) developed this procedure. In analogous terms, you could describe ES as an optimization method that simulates the transition of generations through asexual reproduction, mutation and selection; in contrast, GA simulates sexual reproduction employing crossovers to get recombination.

Schiex and Gaspin (Schiex & Gaspin, 1997; Gaspin & Schiex, 1997) created a GA for genetic mapping. Their so-called *crossover operator* generates a child candidate solution by recombining the orders of two parent candidate solutions. Their *mutation operator* is a simple position exchange of two loci, which is followed by a systematic order reshuffling. With an efficient EM algorithm implementation, each order is evaluated for its multipoint maximum likelihood in the reshuffling procedure.

Mester *et al.* (2003a) developed an ES mapping algorithm in which the mutation stage consists of several steps. The first is a change in the order, possibly at multiple loci. This is followed by three types of systematic order reshuffling. The initial set of candidate solutions is generated by applying this reshuffling five times to random orders. The optimization criterion is the sum of adjacent distances or recombination frequencies. Mester *et al.* (2003b) enhanced the algorithm to be able to deal with F_2s in which dominantly observed loci segregated in repulsion phase. For this purpose, the loci are split into two sets, each having only dominantly observed loci in coupling phase as well as the codominantly observed loci. Thus, dominantly observed loci of the two sets are in repulsion phase with each other, whilst the codominant loci are in both sets. In the enhanced ES algorithm, the sum of adjacent distances in both maps is optimized under the restriction of identical orders of the codominant loci. For this system to work, you need at least two codominantly observed loci and preferably more. As a final step, the two separate maps are integrated by taking the average distance between codominant neighbours and interpolating the dominant loci. The same problem was also treated by Jansen (2009) using simulated annealing in conjunction with multipoint imputation of missing data.

7.8 Ant colony optimization

The way a colony of ants finds an efficient path from its nest to a source of food, inspired Dorigo (1992) and colleagues (Colorni *et al.*, 1992) to yet another optimization algorithm. It appears that, in nature, ants are relatively successful in finding the shortest route. They do this by depositing a chemical substance, called a pheromone, as scent on the route they take. Ants new to the route simply follow the trail with the strongest scent. Iwata & Ninomiya (2006) applied the algorithm to the genetic mapping problem. In the basic algorithm, an ant starts in a random locus and goes to the next locus with a probability that depends on both the distance to the locus (inversely) and the amount of pheromone accumulated (by previously passing ants) on the path towards it. As a distance measurement, the pairwise recombination frequency or the pairwise log-likelihood is used, and thus the optimization criterion is the sum of adjacent recombination frequencies or log-likelihoods. After completion, each ant drops a fixed amount of pheromone on the route taken, which results in short routes obtaining much pheromone per segment whereas long routes get just a little. To adapt the algorithm from being circular (like the travelling salesman), a similar system is applied to loci. The start locus and end locus of the path receive pheromone, whilst they become selected with a probability that depends on the accumulated amount of pheromone. All these steps are repeated for a given number of ants (e.g. 20). It is only after these ants have finished that they drop their pheromone on the paths and on the start and end locus they have taken. Pheromones are volatile substances that evaporate at a given rate. From the newly dropped pheromones and the evaporation rate, the pheromone levels of the loci and the paths are updated. Repeatedly, series of ants are sent to find a route according to these rules (e.g. 5000 series). To prevent ending up at local optima, the algorithm had to be modified by introducing extra randomness whilst on the other hand placing more weight on the best routes. The algorithm does have quite a large number of parameters: (i) the initial amount of pheromone for each locus and (ii) for each segment, (iii) the amount of pheromone dropped by each ant, (iv) the pheromone evaporation rate, (v) the number of ants per series, (vi) the number of series, (vii) the relative influence of the pheromone in choosing the start locus and on each path to the next locus, (viii) the rate of extra randomness, and (ix) the weight placed on the best routes. For the inexperienced user, selecting the proper settings might not be easy.

7.9   Distance geometry

The final algorithm we address is the algorithm of Newell *et al.* (1995), which uses distance geometry to find the optimal map. It is not an optimization algorithm that is used to solve the travelling salesman problem. The algorithm explores the high-dimensional space of the loci but in a way that is different from the MST algorithm of Wu *et al.* (2008) (see Section 7.4). First, a matrix of all pairwise map distances (i.e. in centiMorgans) and a matrix of their variances are created. Next, the eigenvectors and eigenvalues are determined. As was mentioned above, the loci of a linkage group should lie in a one-dimensional space, whilst missing and erroneous observations will create more dimensions. Thus, the first principal axis (i.e. that of the largest eigenvalue) becomes the backbone of the map onto which all the points can be projected to obtain the final map. Studying the two- or three-dimensional coordinates can be used to identify problematic loci, if present.

7.10   Final remarks

At present, it is not quite crystallized which of the methods are the best, in terms of (i) quality of the outcome, (ii) speed and (iii) applicability to all genetic situations (i.e. types of segregating population, numbers of loci). The way in which a method is implemented in computer software plays an important role in all three aspects. Molecular marker techniques have evolved greatly over the last decade, and the numbers of markers observed in the segregating populations are becoming larger and larger. Due to this, the speed aspect will become a major factor in the choice of algorithm, but it should not surpass the quality aspect.

Exercise

In the set of data files available for downloading at the Cambridge University Press web site at http://www.cambridge.org/9781107013216, there are eight files with the data for four simulated populations. The populations are of the backcross (BC_1) type and of the F_2 type, each type with 101 and 401 loci. Each population consists of 100 individuals. The simulations were performed without crossover interference, so that the Haldane mapping function applies. The names of the loci correspond to their position on the simulated map. All loci were distributed evenly over a single chromosome of 100 cM length. The locus genotypes are given for each population with no observations missing and without errors. Additionally, the same populations are given for which a random 5% of the observations was made and also a random 0.5% of the observations was changed to an alternative and thus erroneous genotype. The data files with the locus genotypes are given as tab-delimited text files, which can easily be converted into the specific formats of the different mapping software. The genotype codes used are: a like the first parent, h like the F_1 and b like the second parent, and '$-$' is a missing observation. The corresponding file names are given in the following table:

Population type	Missing, error	101 Loci	401 Loci
BC_1	0%, 0%	BC1_101.txt	BC1_401.txt
BC_1	5%, 0.5%	BC1_101_5%m_0.5%e.txt	BC1_401_5%m_0.5%e.txt
F_2	0%, 0%	F2_101.txt	F2_401.txt
F_2	5%, 0.5%	F2_101_5%m_0.5%e.txt	F2_401_5%m_0.5%e.txt

Several of the methods described in this chapter come with software that is freely available over the internet. Compare the performance of the mapping algorithms on these datasets for the software packages that you have access to. Compare both the speed and the final resulting orders. Compare estimated orders with the true order with the ranks of the loci in the orders using the following measures. Given n loci, and locus i (with true position i) obtained position p_i in the estimated order, then the '*restoration quality*' (Mester *et al.*, 2003a) is:

$$K_r = (n - 1) \Big/ \sum_{i=1,\,n-1} |p_i - p_{i+1}|,$$

Spearman's rank correlation coefficient is:

$$\rho_s = 1 - 6 \sum_{i=1,\,n} (i - p_i)^2 / (n^3 - n),$$

and Kendall's rank correlation coefficient is:

$$\tau = -1 + 4C/(n^2 - n),$$

in which C is the number of concordant pairs of true with obtained positions. Two pairs, (i, p_i) and (j, p_j), are said to be concordant if $\mathrm{sgn}(i - j) = \mathrm{sgn}(p_i - p_j)$.

Because the estimated order is equivalent to its reverse, we should consider the absolute values of the correlation coefficients.

References

Aarts, J. K. & Lenstra, E. L. (2003). *Local Search in Combinatorial Optimization*. Princeton, NJ: Princeton University Press.

Applegate, D. L., Bixby, R. E., Chvátal, V. & Cook, W. J. (2007). *The Traveling Salesman Problem: a Computational Study*. Princeton, NJ: Princeton University Press.

Buetow, K. H. & Chakravarti, A. (1987). Multipoint gene mapping using seriation. I. General methods. *American Journal of Human Genetics*, 41, 180–8.

Cartwright, D. A., Troggio, M., Velasco, R. & Gutin, A. (2007). Genetic mapping in the presence of genotyping errors. *Genetics*, 176, 2521–7.

Colorni, A., Dorigo, M. & Maniezzo, V. (1992). An investigation of some properties of an ant algorithm. In *Proceedings of the Second International Conference on Parallel Problem Solving from Nature*, eds. Männer, R. & Manderick, B. Amsterdam, The Netherlands: Elsevier, 509–20.

Dorigo, M. (1992). *Optimization, Learning and Natural Algorithms*. PhD thesis, Politecnico di Milano, Italy.

Ellis, T. H. N. (1999). Neighbour mapping as a method for ordering genetic markers. *Genetics Research*, 69, 35–43.

Gaspin, C. & Schiex, T. (1997). Genetic algorithms for genetic mapping. In *Proceedings of Evolution Artificielle 97*. Nîmes, France.

Glover, F. (1986). Future paths for integer programming and links to artificial intelligence. *Computers and Operations Research*, 13, 533–49.

Hackett, C. A., Pande, B. & Bryan, G. J. (2003). Constructing linkage maps in autotetraploid species using simulated annealing. *Theoretical and Applied Genetics*, 106, 1107–15.

Holland, J. H. (1975). *Adaptation in Natural and Artificial Systems*. Ann Arbor, MI: University of Michigan Press.

Iwata, H. & Ninomiya, S. (2006). AntMap: constructing genetic linkage maps using an ant colony optimization algorithm. *Breeding Science*, 56, 371–7.

Jansen, J. (2005). Construction of linkage maps in full-sib families of diploid outbreeding species by minimizing the number of recombinations in hidden inheritance vectors. *Genetics*, 170, 2013–25.

Jansen, J. (2009). Ordering dominant markers in F2 populations. *Euphytica*, 165, 401–17.

Jansen, J., De Jong, A. G. & Van Ooijen, J. W. (2001). Constructing dense genetic linkage maps. *Theoretical and Applied Genetics*, 102, 1113–22.

Kirkpatrick, S., Gelatt, C. D. Jr & Vecchi, M. P. (1983). Optimization by simulated annealing. *Science*, 220, 671–80.

Mester, D., Ronin, Y., Minkov, D., Nevo, E. & Korol, A. (2003a). Constructing large scale genetic maps using evolutionary strategy algorithm. *Genetics*, 165, 2269–82.

Mester, D., Ronin, Y. I., Hu, Y., Peng, J., Nevo, E. & Korol, A. B. (2003b). Efficient multipoint mapping: making use of dominant repulsion-phase markers. *Theoretical and Applied Genetics*, 107, 1102–12.

Newell, W. R., Mott, R., Beck, S. & Lehrach, H. (1995). Construction of genetic maps using distance geometry. *Genomics* 30, 59–70. [Erratum *Genomics*, 34, 283, 1996.]

Rechenberg, I. (1973). *Evolutionsstrategie: Optimierung technischer Systeme nach Prinzipien der biologischen Evolution*. Stuttgart, Germany: Fromman-Holzboog.

Schiex, T. & Gaspin, C. (1997). CarthaGene: constructing and joining maximum likelihood genetic maps. In *Proceedings of the Fifth International Conference on Intelligent Systems for Molecular Biology 97*. Halkidiki, Greece.

Stam, P. (1993). Construction of integrated genetic linkage maps by means of a new computer package: JoinMap. *Plant Journal*, 3, 739–44.

Tan, Y. D. & Fu, Y. X. (2006). A novel method for estimating linkage maps. *Genetics*, 173, 2383–90.

Van Os, H., Stam, P., Visser, R. G. F. & Van Eck, H. J. (2005). RECORD: a novel method for ordering loci on a genetic linkage map. *Theoretical and Applied Genetics*, 112, 30–40.

Weeks, D. E. & Lange, K. (1987). Preliminary ranking procedures for multilocus ordering. *Genomics*, 1, 236–42.

Wu, Y., Bhat, P., Close, T. J. & Lonardi, S. (2008). Efficient and accurate construction of genetic linkage maps from the minimum spanning tree of a graph. *PLoS Genetics*, 4, e1000212.

8 Outbreeding species

The genetic circumstances found in outbreeding species are more complicated than in inbreeding species: more than two alleles may be present at each locus and the linkage phases may vary across loci and between the parents of an experimental cross. The experimental design of inbreeding species often cannot be applied to outbreeding species. This chapter focuses on explaining in detail the genetic situations encountered in outbreeding species. It further describes the linkage analysis of a full-sib family of an outbreeding species.

8.1 Introduction

In the preceding chapters, we treated linkage mapping for experimental populations derived from a cross between two fully homozygous parents. Homozygosity is obtained by many generations of self-fertilization or sib mating, or by creating doubled haploids. For many plant species, self-pollination followed by self-fertilization (also called *autogamy*) is the normal mode of sexual reproduction. In some *hermaphrodite* species (i.e. species with organs of both sexes), a system of self-incompatibility or self-sterility causes an obstruction to self-fertilization. In *dioecious* species (dioecious means that individuals have organs of only one of the two sexes), self-fertilization is of course impossible. In outbreeding species, for which the normal mode of sexual reproduction is by crossing with other individuals (also called *allogamy*), forced inbreeding usually results in individuals with a poor viability. This phenomenon is called inbreeding depression. The consequence for such species is that linkage analysis can often only be performed using populations obtained by crossing relatively unrelated individuals.

Two important effects of outbreeding as the normal mode of sexual reproduction are that individuals are heterozygous at many loci and that natural populations are heterogeneous, i.e. there are no identical genotypes. Therefore, linkage analysis of populations produced by outbreeding has to take into account the fact that the number of alleles segregating in the population may vary between loci and that the linkage phases are unknown. In contrast, linkage analysis of populations derived from two fully homozygous parents has to deal with two segregating alleles for each locus, whilst the linkage phases are known.

Many outbreeding species can produce only a limited number of progeny for each (female) parent, as, for instance, in many domestic animal species. In such situations, many families and pedigrees have to be combined in order to obtain sufficient offspring. This adds complexity to the analysis, because the amount of information about the transmission of alleles and linkage phases is importantly less than when you are dealing with a single, large full-sib family. Linkage analysis based on a set of many small families and pedigrees is beyond the scope of this book. There are also many outbreeding species for which it is possible to generate a single full-sib family that is sufficiently large for a successful linkage analysis. Examples are tree species (apple, citrus, grape, eucalyptus, oil palm, pine, willow), grasses (miscanthus), fish species (cod, guppy, salmon), amphibians (frog), crustaceans (crab, prawn) and molluscs (abalone, oyster). This chapter addresses the specifics of linkage analysis based on a single full-sib family of an outbreeding species. Linkage analysis of inbreeding species also concerns a single full-sib family (i.e. the BC_1 and F_2 are strictly full-sib families, whilst others, like RILs, may be regarded more loosely as full sibs due to their two common founding ancestors). A major distinction between the in- and outbreeding species full-sib families concerns the number of alleles that segregate in the offspring of the cross. For inbreeding species, this is limited to two alleles, whereas for outbreeding species, up to four alleles may segregate, whilst in addition this may vary across the loci from two to four. Another distinction is that segregation is already present in the first generation after the cross with an outbreeding species, whereas after crossing homozygous lines, it is only the second generation and later that reveal segregation.

8.2 Segregation type and linkage phases

To explain the more complex segregation in outbreeding species, we use the concept of *segregation type* (Maliepaard *et al.*, 1997). A segregation type is a description of the parental genotypes at a locus, for instance $ab \times cd$. The two characters on the left of the '×' represent the alleles of the first parent and the two characters on the right those of the second parent. Each different character represents a distinct allele. Furthermore, the position of each character in a pair defines its *grand*parental origin. For instance, below is the scheme for the genotypes at a locus with segregation type $ab \times cd$:

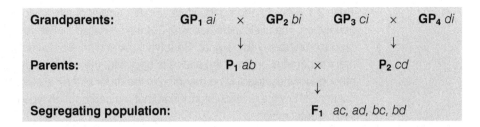

In correspondence with the segregation type, the grandparents GP_1 and GP_2 passed on the alleles a and b, respectively, to parent P_1, whilst the grandparents GP_3 and GP_4 passed on the alleles c and d, respectively, to parent P_2. The second alleles of the grandparents, all indicated by i, are irrelevant for now. The parents P_1 and P_2, with genotypes ab and cd, are crossed to produce offspring with possible genotypes ac, ad, bc and bd. Thus, conveniently, the segregation type also defines the genotypes of the offspring. Finally, it is essential to know the genotypes of the parents to be able to determine the segregation type.

Ritter & Salamini (1996), Maliepaard *et al.* (1997) and Wu *et al.* (2002a) distinguished seven essentially different segregation types, which are shown in Table 8.1. Table 8.2 presents a list of other genetically equivalent segregation types. Different characters are used for the alleles in the different segregation types and their offspring genotypes, because this allows the implicit recognition of the corresponding segregation type when dealing with a given genotype. For example, the segregation type of genotype np is instantly recognized as being $nn \times np$. We make an exception for the allele indicated with 0, which is present in three segregation types. This is the so-called *null allele*, which does not produce an observable signal. The presence of the null allele gives rise to *dominance*, which is the situation in which a heterozygous genotype cannot be distinguished from a homozygous genotype. For instance, the heterozygous genotype $q0$ is phenotypically the same as the homozygous genotype qq. Some molecular marker systems allow the detection of allele dosage, which does enable the phenotypic distinction between qq and $q0$ (or $0q$), between rr and $r0$, and between tt and $t0$. In such instances, another segregation type will be applicable, i.e. $q0 \times q0$ becomes effectively $hk \times hk$, whilst $rs \times r0$ and $t0 \times tu$ become effectively $ab \times cd$. Figure 8.1 illustrates the configuration of the alleles on the homologous chromosomes in the parents for the seven distinct segregation types.

In many research experiments, the knowledge of the grandparental genotypes is lacking. Even if this knowledge is available, the grandparental genotypes may have identical alleles, so that it is not possible to know the grandparental origin of the alleles in the parents (note that this puts restrictions on the four alleles i in the scheme illustrating the $ab \times cd$ segregation type at the start of this section).

Table 8.1. Distinct segregation types, their number of alleles and their offspring genotypes. The allele indicated with *0* is the so-called *null allele*, which does not produce an observable signal. Genotypes *hk* and *kh* are distinct here to indicate that the parental origin of the alleles is opposite; the same holds for *q0* and *0q*. In other genotypes, there is no meaning in the order of their alleles. The last two columns show the level of information that the segregation type provides about the two parental meioses.

Segregation type	No. of alleles	Offspring genotypes	Genotypes with identical phenotypes	Information of	
				P_1	P_2
ab×*cd*	4	*ab, ac, bc, bd*		Complete	Complete
hk×*hk*	2	*hh, hk, kh, kk*	*hk, kh*	Partial	Partial
lm×*ll*	2	*ll, lm*		Complete	None
nn×*np*	2	*nn, np*		None	Complete
q0×*q0*	2	*qq, q0, 0q, 00*	*qq, q0, 0q*	Partial	Partial
rs×*r0*	3	*rr, r0, rs, s0*	*rr, r0*	Complete	Partial
t0×*tu*	3	*tt, tu, t0, u0*	*tt, t0*	Partial	Complete

Table 8.2. Segregation types that are genetically equivalent to the types in Table 8.1.

Equivalent to:		
ab×*cd*	*lm*×*ll*	*nn*×*np*
ab×*ac*	*ab*×*cc*	*aa*×*bc*
ab×*c0*	*a0*×*bb*	*aa*×*b0*
a0×*bc*	*ab*×*00*	*00*×*ab*
a0×*b0*	*a0*×*00*	*00*×*a0*

How do we deal with such conditions? Suppose, for instance, we know that the parental genotypes of the cross for a certain locus are *ab* and *cd*, i.e. there are four distinct alleles. We are supposed to define the allele coming from the first grandparent as the *a* allele, but unfortunately, we do not know the grandparental origin of the alleles. To be able to deal with such situations, we use the concept of the *phase type*. The phase type describes the linkage phase situation in both parents using two digits that can be 0 or 1, with one digit for each parent. In

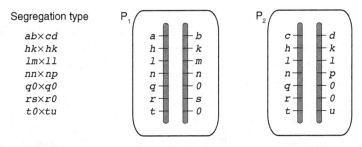

Segregation type

$ab \times cd$	
$hk \times hk$	
$lm \times ll$	
$nn \times np$	
$q0 \times q0$	
$rs \times r0$	
$t0 \times tu$	

Fig. 8.1

Configuration of the alleles on the two homologous chromosomes in the two parents for the seven distinct segregation types in Table 8.1.

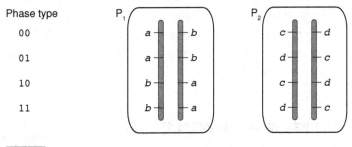

Phase type

00

01

10

11

Fig. 8.2

Configuration of the alleles on the two homologous chromosomes in the two parents for the segregation type $ab \times cd$ for all four phase types.

essence, the linkage phase configuration distinguishes the grandparental origin, i.e. whether the alleles are on the same haplotype or not. If the phase type digit equals 0, then the origin of the alleles is according to the definition of the segregation type. If the phase type digit equals 1, then the origin of the alleles is opposite to the definition of the segregation type. Figure 8.2 illustrates the allelic configurations of segregation type $ab \times cd$ for the four possible phase types. As we shall see later, by comparing pairwise recombination frequency estimates under all linkage phases (i.e. each phase setting has its particular estimator), it is usually possible to distinguish the linkage phases in both parents. By using the appropriate phase type, it thus becomes possible to swap the grandparental origin of the alleles. For example, suppose, we have an $ab \times cd$ locus **L1**, for which the grandparental allelic origins are according to the definition (i.e. 00), and another $ab \times cd$ locus **L2**, for which the grandparental allelic origins are unknown. Suppose, after computing and comparing the recombination frequencies under all phase settings, it turns out that the loci are in coupling phase in P_1 and in repulsion phase in P_2. In our coding system, this means that the *a* allele of **L1** is in coupling phase with the *a* allele of **L2** (as are their *b* alleles), whilst the *c* allele of **L1** is in repulsion phase with the *c* allele of **L2** (as are their *d* alleles). Then, by setting the phase type of **L2** to 01, we swap the grandparental origin of its alleles *c* and *d* (due to the 1 in the phase type), whilst leaving those of its alleles *a* and *b* according to the definition of the segregation type (due to the 0 in the phase type). The consequence of the system is that we can assign the alleles according to the segregation type without knowing the grandparental origin, because that will be resolved when estimating the pairwise recombination frequencies. Now, do we need at least a single locus with known grandparental allelic origins? The answer is no, because it is irrelevant, which of the two grandparents of P_1 is labelled the first and which the second grandparent, so long as it is identical for all loci studied, i.e. of the same linkage group. The same holds for grandparents three and four with respect to P_2. In other words, for linkage analysis, it is not important which of the two homologous chromosomes comes first and which second in both parents in Fig. 8.2. Of course, when selecting for certain alleles of traits after their linkage analysis, it is important to take the phase types into account.

8.3   Two-way pseudo-testcross strategy

In linkage analysis, essentially we try to detect the occurrence of recombination events between loci. For diploid species, recombination occurs in the two parents. The different segregation types of the outbreeding species full-sib family allow the observation of recombination separately for both parental meioses. Table 8.1 shows that the segregation types vary with respect to the level of information they provide regarding the parental meioses. The *ab×cd* segregation type provides complete information on both parents. The *lm×ll* and *nn×np* types only provide information on the meiosis of the first or second parent, respectively. The *hk×hk* type provides information on both parental meioses, but only partially, because for a heterozygous offspring phenotype *hk*, it is uncertain whether the first parent provided the *h* allele and the second parent the *k* allele, or the other way around. The *q0×q0* type provides even less information than the *hk×hk* type because of dominance. The *rs×r0* and *t0×tu* types have reduced information due to dominance in one of the parents.

   Why can we not get such separated information in crosses with inbreeding species? The F_2 and the RIL family types are generated by selfing the F_1 and for RILs the subsequent generations. In this scheme, the same individuals act as both parents and this fact hampers the distinction between the two parental meioses. The scheme is similar to the *hk×hk* (and *q0×q0*) segregation type, with the particular detail that the linkage phases are identical in both parents, i.e. the F_1. The experimental population types based on backcrossing to homozygous parents do not provide information on the backcross parent, because the homozygous parent's loci do not segregate. By using reciprocal backcross populations, it is possible to study differences in recombination between the maternal and paternal meiosis in inbreeding species.

As a way to circumvent the complexities of analysing two meioses simultaneously in the outbreeding species full-sib family, Grattapaglia & Sederoff (1994) proposed the so-called *two-way pseudo-testcross* strategy. Most of the molecular markers they had access to (they used random amplified polymorphic DNA markers, or RAPDs), were of the *lm×ll* and *nn×np* types. They did not have many markers with a more informative segregation type, which can act as bridges in a combined analysis of both parental meioses. The few bridge markers they had, they used for establishing homology between linkage groups of the separate parental sets.

How exactly does the two-way pseudo-testcross strategy proceed? We have to make a separation of the genotypes into parental datasets, i.e. one set contains all, and only, the *lm×ll* type loci, whilst the other has the *nn×np* type loci. Loci of the other segregation types may be added to both sets after recoding their genotypes according to the *lm×ll* and *nn×np* segregation types. For the recoding, we follow the property that the positions of the alleles in a segregation type define their grandparental origin. The genotypes *ll* and *lm* both received an *l* allele from the second parent. Likewise, both *nn* and *np* genotypes receive an *n* allele from the first parent. Thus, only the second of the two characters of the genotype code is relevant. For the recoding towards the first parent, it is relevant for the *ab×cd* genotypes whether the *a* or the *b* allele was transmitted. These are equivalent to the *l* and *m* alleles, respectively. Thus, genotypes *ac* and *ad* must be recoded as *ll*, whilst *bc* and *bd* must be recoded as *lm*. For the recoding towards the second parent, the transmission of the *c* or *d* allele is relevant. These are equivalent to the *n* and *p* alleles, respectively. Thus, genotypes *ac* and *bc* must be recoded as *nn*, whilst *ad* and *bd* must be recoded as *np*. For the other segregation types, the recoding is done in a similar way; all recodings are given in Table 8.3. For several phenotypes the translation is impossible and results in the 'unknown' genotype '- -'. For phenotype *hk*, it is unresolved which parent transmitted which allele. For the phenotypes comprising genotypes *rr* and *r0*, it is impossible to determine whether the second parent transmitted the *r* or the *0* allele. Similarly, for the phenotypes comprising genotypes *tt* and *t0*, the transmission of the first parent's alleles cannot be resolved. As a result, only half of the individuals of these segregation types (*hk×hk*, *rs×r0*, *t0×tu*) are expected (according to Mendelian ratios) to be available in the analyses. For segregation type *q0×q0*, only the *00* genotype will

Table 8.3. Recoding of genotypes for separate parental datasets. There is no meaning in the order of the alleles in the genotypes, i.e. $hk = kh$, $q0 = 0q$. The code '- -' is for the unknown genotype.

Original genotype	Genotype in $lm \times ll$	Genotype in $nn \times np$	Original genotype	Genotype in $lm \times ll$	Genotype in $nn \times np$
$ab \times cd$			$rs \times r0$		
ac	ll	nn	$rr, r0$	ll	$- -$
ad	ll	np	rs	lm	nn
bc	lm	nn	$s0$	lm	np
bd	lm	np			
$hk \times hk$			$t0 \times tu$		
hh	ll	nn	$tt, t0$	$- -$	nn
hk	$- -$	$- -$	tu	ll	np
kk	lm	np	$u0$	lm	np
$q0 \times q0$					
$qq, q0$	$- -$	$- -$			
00	lm	np			

translate to the lm and np genotypes. Its other genotypes, being indistinct as a phenotype, can only be translated to unknowns. Because of this, only a quarter of its individuals are expected to be informative. Therefore, loci with this segregation type are not very useful in these analyses.

Linkage phase determination in the pseudo-testcross

Grattapaglia & Sederoff (1994) used the term *pseudo* in pseudo-testcross to indicate the difference from the regular testcross, in which the parent with which it is crossed is always homozygous. A difference with the regular inbreeding species backcross is that the linkage phases may be unknown beforehand, whereas in the BC_1, the knowledge of the genotype of the parents of the F_1 reveals all heterozygous loci to be in coupling phase with each other. If the grandparental genotypes are known and sufficiently informative, then the linkage phases may be known beforehand, but this does not occur often in practice. Thus, usually, the linkage phases must be determined for all loci. Formally, it is better to describe the process as an inference rather than a determination. However, we find it may also be called linkage phase determination, because in practice the likelihood of one phase is usually much larger than that of the other.

As we saw in Chapter 3, the estimation of the recombination frequency in the backcross is rather straightforward: tally the recombinant and non-recombinant individuals and divide the recombinant count by the total count. If the linkage phase is opposite from what was assumed, then the recombinant individuals must be classified as non-recombinant and the non-recombinant ones as recombinant. The effect of this is that the estimator of the recombination frequency under the assumption of coupling phase, r_c, is the complementary frequency of that under repulsion phase, r_r:

$$r_c = 1 - r_r.$$

Because the theoretical maximum value of the recombination frequency is $\frac{1}{2}$, the recombination frequency estimate above $\frac{1}{2}$ becomes $\frac{1}{2}$ under maximum likelihood (and thus the *LOD* is 0). The linkage phase is chosen, under which assumption the estimate is below $\frac{1}{2}$ (and has a positive *LOD*).

The maximum of $\frac{1}{2}$ is a theoretical maximum, i.e. it holds under a population of infinite size. In practice, population sizes are always limited. We may regard a segregating population as a finite sample of gametes that are recombinant or non-recombinant for a pair of loci. In the sampling process, random variation can cause the number of recombinants to be larger than the number of non-recombinants, resulting in a recombination frequency estimate that is larger than the theoretical maximum of $\frac{1}{2}$. Normally, these situations occur only when the sample size is small and/or when the recombination probability is close to $\frac{1}{2}$, i.e. loci are far apart on the chromosome. Such circumstances may result in contradictory conclusions. For instance, locus **A** could be concluded to be in coupling phase with **B** and locus **B** in coupling phase with **C**, whereas **A** is in repulsion phase with **C**. The latter fact contradicts the indirect phase relationship through locus **B**, which is coupling phase. In order to prevent problems of this kind, we sort all pairs based on the significance of the association, and start assigning linkage phases from the most significant pairs downwards and stop as soon the linkage phases of all loci are assigned.

If the grandparental origin is known for at least one locus, then the linkage phases of the other loci can be oriented upon this one. A phase type of a single digit (0 or 1) can be used to indicate (non-)correspondence with the grandparental origin of the alleles according to the testcross segregation type. If the grandparental origin is not known for any locus, then any locus can be chosen as the origin.

8.4 Estimation of the pairwise recombination frequency in the full-sib family

The diploid full-sib family is the product of the meioses of their two parents. The probability of recombination can be assumed to be the same in male and female meiosis. For the estimation of the recombination frequency between pairs of loci, their segregation types as well as their linkage phases must be taken into account. Several authors, e.g. Ritter & Salamini (1996) and Maliepaard *et al.* (1997), presented maximum likelihood methods to estimate the pairwise recombination frequency between loci in the outbreeding species full-sib family. They showed that explicit solutions to the likelihood equations can be derived for several combinations of segregation types. For combinations that have no explicit solutions (because they are the roots of higher-order polynomials), Maliepaard *et al.* (1997) presented a general EM algorithm, which can be seen as an extension of the EM algorithm that was presented in Chapter 3 for the F_2. For this method, the likelihood for the full-sib family with respect to the recombination frequency, r, is set up considering the two gametes that generated each genotype. Each genotype may have 0, 1 or 2 recombinations in the two gametes between the two loci considered. If we could observe the complete genotype (i.e. including linkage phases) of each individual in the family, then we could count the number of recombinant gametes, R, and non-recombinant gametes, S, in the family, i.e. under the assumption of certain linkage phases in the parents. Thus, the likelihood (ignoring the multinomial coefficient) is:

$$L(r) \propto (1 - r)^S r^R,$$

so that the maximum likelihood estimator of r becomes (as in Chapter 3):

$$\hat{r} = R/(S + R) = R/(2N), \tag{8.1}$$

in which N is the total number of observed individuals. The problem is that we do not know the genotypes underlying all phenotypes, which means that we cannot classify their gametes with respect to recombination. However, supposing we know r and the linkage phases in the parents, then we can calculate for every phenotype, f, the probability π_{fh} of the underlying haplotype combination (i.e. a set of two gametes) that contains $h = 0$, 1 or 2 recombinations. Subsequently, the number of recombinant gametes for each individual i with phenotype f has the expectation:

$$R_{fi} = (0\pi_{f0} + 1\pi_{f1} + 2\pi_{f2})/(\pi_{f0} + \pi_{f1} + \pi_{f2})$$
$$= (\pi_{f1} + 2\pi_{f2})/(\pi_{f0} + \pi_{f1} + \pi_{f2}).$$

The sum of R_{fi} over all individuals becomes the expectation of R:

$$R = \sum_i R_{fi}.$$

Using $r = \frac{1}{4}$ as the starting value, we can calculate the expectation of R (the E-step), and with this value of R, we can obtain a new value of r with (8.1) (the M-step). With this new value of r, we can improve the expectation of R, with which we can obtain an improved estimate of r, and so on. This EM procedure is repeated until there is no more change in the value of r. This is the recombination frequency estimate under the assumed parental linkage phases, and under the assumption that the probability of recombination is equal for the two parental meioses.

Linkage phase determination in the full-sib family

As stated above, the grandparental origin of the alleles in the parents may be unknown, which makes it necessary to infer the parental linkage phases based on the observations. Because there are two parents, there are also two linkage phases, and because we are dealing with outbreeding species, the two phases need not be identical as they are with the F_2. Thus, there are four linkage phase combinations, coupling (c) or repulsion (r) in two parents, which are indicated by $c \times c$, $c \times r$, $r \times c$ and $r \times r$, i.e. the positions on the left and right of the '$\times$' describe the phase in the first and second parent, respectively. If one or both of the loci are homozygous in one of the parents (i.e. $lm \times ll$, $nn \times np$), then the linkage phase in that parent is undefined. There are estimators for each of the four phase situations: $r_{c \times c}$, $r_{c \times r}$, $r_{r \times c}$ and $r_{r \times r}$ (or just two if the linkage phase in one parent is undefined, whilst the recombination frequency cannot be estimated between an $lm \times ll$ and an $nn \times np$ locus). Maliepaard *et al.* (1997) presented an extensive treatment on how the results of the estimates can be used to infer the linkage phases. Here, we present just the main features. In general, the four estimates will have the following relationship:

$$r_{c \times c} = 1 - r_{r \times r} \quad \text{and} \quad r_{c \times r} = 1 - r_{r \times c}.$$

If the real phase combination is $c \times c$ or $r \times r$, then the estimates $r_{c \times r}$ and $r_{r \times c}$ will have values close to $\frac{1}{2}$ and, inversely, if the real phase combination is $c \times r$ or $r \times c$, then the other estimates are about $\frac{1}{2}$. Of the two estimates that clearly deviate from $\frac{1}{2}$, only the one smaller than $\frac{1}{2}$ is a legitimate estimate under maximum likelihood. Thus, the inferred phase combination is that for which the corresponding estimate has the smallest value.

There are some specific problems with the above procedure concerning the following situations: (i) at least one of the loci is homozygous in one of the parents, (ii) both segregation types are symmetrical (i.e. $hk \times hk$, $q0 \times q0$), and (iii) both segregation types possess a null allele. In situation (i), the linkage phase can only be established for the parent that is heterozygous at both loci. This procedure is similar to that in the pseudo-testcross. In the symmetrical situations in (ii), both estimates $r_{c \times r}$ and $r_{r \times c}$ are always identical, so no distinction can be made regarding these phase combinations. Two loci with segregation type $q0 \times q0$ is the most problematic: only when the true situation is $c \times c$ will the four estimates behave according to the general relationship above, which results in a correct inference of the phase combination. When the true phase combination is one of the other three, then all three estimates under $c \times r$, $r \times c$ and $r \times r$ always result in values less than $\frac{1}{2}$, making it impossible to choose. Situations as in (iii) have similar problems, albeit less extreme; see Maliepaard *et al.* (1997) for details.

In practice, in situations with a mixture of segregation types on one linkage group, the above problems can be circumvented. The linkage phases that remained unresolved based on the rules described above can usually be inferred indirectly through loci with more informative segregation types. For instance, if the phase combination cannot be determined directly between two $q0 \times q0$ type loci **A** and **B**, but if there is a linked $ab \times cd$ type locus **C** with which it is inferred that the phase combination of the pair **AC** is $c \times c$ and of the pair **BC** is $c \times r$, then it can be concluded that the phase combination of the pair **AB** must be $c \times r$.

As with the pseudo-testcross, it is possible to obtain contradictory conclusions regarding the linkage phases. Here, too, problems of this kind are prevented by assigning linkage phases from the most significant pairs downwards and stopping as soon the linkage phases of all loci are assigned.

8.5   Simultaneous estimation of the separate parental recombination frequencies

In the previous section, the pairwise recombination was estimated under the assumption of equal parental recombination probabilities. This assumption is not made in the two-way pseudo-testcross strategy, where the recombination frequency is estimated for each parent separately. In many situations, significant differences exist between the recombination rates in the male and female meioses. In some species, the male meiosis shows a higher level, and in others the female; for overviews, see Burt *et al.* (1991) and Lenormand & Dutheil (2005). Because the meioses of the two sexes are distinct processes, there is no biological reason why recombination rates should be identical. Another factor in this respect is that the full-sib family can be regarded as a combination of independent finite samples of male and female gametes. This will result in random differences in the number of recombination events realized in the two samples, regardless of any systematic differences between the parental meioses. Therefore, a more realistic model for recombination in an outbreeding species full-sib family would be to allow parent-specific recombination frequencies (Wu *et al.*, 2002b). Although theoretically this may also apply to inbreeding species, in practice the usual experimental design (F_2, RIL) does not allow separation of the parental meioses due to the symmetry of the segregation. Consequently, average recombination frequencies are estimated in inbreeding species. (Note that, in the backcross, the recombination is estimated in a single parent, i.e. the F_1. Using the F_1 as both male and female allows separate estimation of recombination frequencies in the maternal and paternal meioses.)

Under a model of parent-specific recombination probabilities, we can define r_1 and r_2 as the recombination frequencies between two loci in the two meioses. Supposing, like before, that we are able to observe the complete genotype of each individual in the family, then we can count the number of recombinant and non-recombinant gametes for the two loci in the two meioses, R_1, R_2, S_1 and S_2, respectively, i.e. under the assumption that the linkage phases in the parents are given. Because the two meioses are independent, we can write the likelihood as (ignoring the multinomial coefficient):

$$L(r_1, r_2) \propto (1 - r_1)^{S_1} r_1^{R_1} (1 - r_2)^{S_2} r_2^{R_2}.$$

The corresponding log-likelihood ℓ is:

$$\ell = \ln(L(r_1, r_2)) \propto S_1 \ln(1 - r_1) + R_1 \ln(r_1) + S_2 \ln(1 - r_2) + R_2 \ln(r_2).$$

The partial derivatives of ℓ towards r_1 and r_2 do not depend on each other, so the maximum likelihood estimators of r_1 and r_2 become (as in Chapter 3):

$$\hat{r}_1 = R_1/(S_1 + R_1) = R_1/N \text{ and}$$
$$\hat{r}_2 = R_2/(S_2 + R_2) = R_2/N,$$

with N being the total number of individuals ($N = R_1 + S_1 = R_2 + S_2$).

Obviously, it depends on the informativeness of the segregation types of the two loci involved (Table 8.1) as to whether it is possible to estimate the two recombination frequencies. For a pair of $ab \times cd$ loci, it is clear that all required observations can be made. If one of the loci in a pair segregates in only one parent, i.e. has segregation type $lm \times ll$ or $nn \times np$, then only the recombination of the corresponding parent can be observed. In this case, the model must be reduced to having only one of the parental recombination frequencies. With other segregation types, some of the two-locus phenotypes will contain multiple (phase-known)

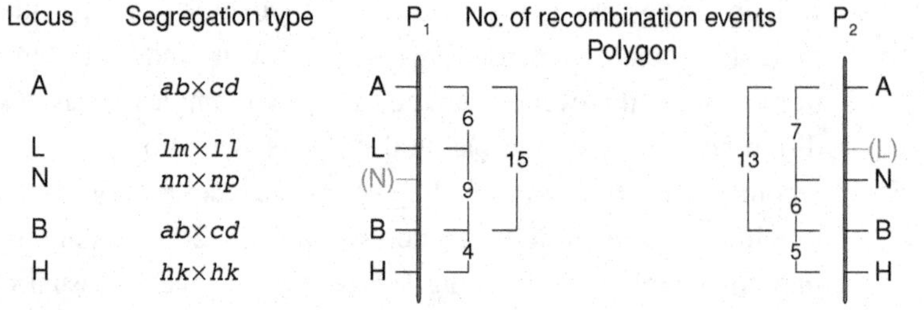

Fig. 8.3

Example of realized recombination events in the two parents for five loci of a mixed set of segregation types.

genotypes. In such cases, estimates can be obtained using an EM procedure as described in Section 8.4 and Chapter 3. The major difference is that it concerns two estimators simultaneously here. Consequently, the expectations in the E-step are expressions in both parameters and therefore the iterations for both estimators must be executed simultaneously. For example, an individual of phenotype ac,hk, under coupling phase in both parents, represents the two phase-known genotypes: ak/ch and ah/ck. Genotype ak/ch is obtained after recombination in P_1 and no recombination in P_2; therefore, its probability is $\frac{1}{4} r_1(1 - r_2)$. Genotype ah/ck is the inverse situation and has probability $\frac{1}{4} (1 - r_1)r_2$. Thus, the contributions of this individual to the expected numbers of recombinant gametes of the first and second parent are, respectively:

$$r_1(1 - r_2)/(r_1(1 - r_2) + (1 - r_1)r_2) \quad \text{and}$$
$$(1 - r_1)r_2/(r_1(1 - r_2) + (1 - r_1)r_2).$$

These expectations are expressions in both parameters, r_1 and r_2. To obtain both estimates, we calculate in each iteration the expectations using the current values of both parameters and maximize for both parameters using these current expectations.

Example

To illustrate the situation of an outbreeding species full-sib family, we have drawn an example with five loci of different segregation types on a chromosome (Fig. 8.3). In the example, the realized recombination events are given for the segments in the two parental meioses in a population of 100 individuals. The situation for the pair of $ab \times cd$ loci **A** and **B** is straightforward: because their segregation types are fully informative, all recombination events can be observed. The separate recombination frequency estimates are: $r_1 = 0.15$ and $r_2 = 0.13$. Under the assumption of equal parental recombination probabilities, we would estimate the average: $r = 0.14$.

The $lm \times ll$ locus **L** segregates only in the first parent, and thus no recombination can be observed in the second parent in any pair with this marker. In other words, locus **L** is completely informative in the first parent and totally *un*informative in the second (Table 8.1). Recombination events can be counted in P_1 between **A** and **L**, and between **L** and **B**, as 6 and 9, respectively (adding up to the 15 between **A** and **B**, so no double recombination events). The pairwise recombination frequency estimates of **A** and **B** with **L** are based on the first parent only, regardless of whether or not we make the assumption of equal parental recombination probabilities: $r_1 = r = 0.06$ for **L–A** and $r_1 = r = 0.09$ for **L–B**. For the $nn \times np$ locus **N**, the situation is the mirror image of **L**: the pairwise recombination frequency estimates of **A** and **B** with **N** are based on only the second parent. It is important to note that it is impossible to observe any recombination event between **L** and **N**, as there is no meiosis in which both markers segregate. Thus, no recombination frequency can be estimated between them directly.

The $hk \times hk$ locus **H** provides partial information: due to its symmetric segregation, it is impossible to designate the parental origin of alleles of the heterozygous offspring hk. As a consequence the recombination frequency between **H** and **B** (as well as between **H** and **A**) can be estimated only under the assumptions of our model of recombination. This means that the fractions of genotypes in coupling and repulsion phase for the offspring phenotype hk are assumed to behave according to the recombination probabilities. For example, the phenotype ac,hk represents the two phase-known genotypes: ak/ch and ah/ck. Under coupling phase in both parents, the first genotype represents a recombination event in the first parent and no recombination event in the second, whereas the second genotype represents the inverse situation. Their probabilities are $\frac{1}{4} r_1(1 - r_2)$ and $\frac{1}{4} (1 - r_1)r_2$, respectively, which are employed to calculate the expectations in the EM procedure as described above. Consequently, the estimates of the separate parental recombination frequencies will be somewhat different from 0.04 and 0.05, respectively. These two values would have been found if locus **H** had been of segregation type $ab \times cd$. The difference depends on how much the realized numbers of (phase-known) genotypes deviate from their expectations. Under the assumption of equal parental recombination probabilities, we would estimate a value close to the average: $r = 0.045$. Here too, the heterozygotes hk are divided over the two phases according to their probabilities $\frac{1}{4} r(1 - r)$ and $\frac{1}{4} (1 - r)r$, and this division is likely to differ slightly from the realization.

For the pairs of locus **H** with locus **L** and locus **H** with locus **N**, the characteristics of the segregation types lead to estimates of the recombination frequencies that represent segregation in a single parent only and, moreover, that are based on only the homozygous observations for locus **H**. For instance, the phenotype *ll,hk* represents the two phase-known genotypes: *lk/lh* and *lh/lk*. Under coupling phase in the first parents, the first genotype represents a recombination event in the first parent, whilst the second genotype represents no recombination event. Their probabilities are $\frac{1}{4} r_1$ and $\frac{1}{4} (1 - r_1)$, respectively, which means that the phenotype class *ll,hk* is expected to consist of a fraction of recombination events that is exactly equal to the estimated recombination frequency. This fact renders the phenotype completely uninformative. Similar derivations can be made for all phenotypes that are heterozygous for locus **H** combined with any phenotype of **L** or **N**. The other phenotype classes consist either of a single recombination event or of no recombination event in the relevant parent, which results in an explicit (non-iterative) estimator of the recombination frequency. Because on average half of the individuals are expected to be heterozygous for locus **H**, the estimates will be based on about half the population only.

Multipoint estimation

In Chapter 5, we mentioned that multipoint estimation of recombination frequencies are more accurate, because the segregation information of all neighbouring loci is being taken into account. Van Ooijen (2011) described an extension of the Gibbs sampling procedure of Jansen *et al.* (2001), which leads to multipoint maximum likelihood estimates of the recombination frequencies, to the outbreeding full-sib family under a model of parent-specific recombination probabilities. Gibbs sampling is a method for the imputation of missing information. For instance, for the *hk* phenotypes, the linkage phase is the missing information. In the presence of loci with completely informative segregation types regarding one or both meioses on the same linkage group, the Gibbs sampler manages to resolve much of this missing phase information. For example, suppose for three loci in the order **L1–L2–L3** an individual has the phenotypic observation *ac–hk–ad*. We assume coupling phases in both parents, and thus there are two optional phase-known genotypes: *aha/ckd* and *aka/chd*. The phenotype implies that, between **L1** and **L2**, depending on the phase orientation of the **L2** phenotype *hk*, there is either a recombination event in the first parent's meiosis or one in the second parent's meiosis, i.e. phase-known genotypes *ak/ch* and *ah/ck*, respectively. These configurations have the roughly equal probabilities $\frac{1}{4} r_1(1 - r_2)$ and $\frac{1}{4} (1 - r_1)r_2$, if studying only the pair **L1** and **L2** (r_1 and r_2 apply to segment **L1–L2**). If studying the pair **L2** and **L3**, the two phase orientations of **L2** correspond to either a recombination event in each parent's meiosis or no recombination event, i.e. phase-known genotypes *ka/hd* and *ha/kd*, respectively. Their (two-point) probabilities are $\frac{1}{4} r_1 r_2$ and

$\frac{1}{4}(1 - r_1)(1 - r_2)$, respectively, thereby making the second configuration much more likely (here, r_1 and r_2 apply to segment **L2–L3**). In the multipoint analysis this relationship of **L2** with **L3** affects its relationship with **L1**, because it makes the recombination event between **L1** and **L2** much more likely to occur in the second parent's meiosis (*aha/ckd*) than in the first parent's (*aka/chd*). In this way, it will affect the estimation of the separate parental recombination frequencies of the two segments, thereby improving accuracy of the result. This example is just a three-point situation, which could be dealt with in an analytical way; however, the multipoint approach works in a similar fashion over longer stretches of loci with various types of missing information in the observations.

Currently, increasing use is made of single-nucleotide polymorphism (SNP) markers, because they appear to possess various practical (and possibly financial) advantages over other types of markers. SNPs usually have just two alleles and therefore using SNPs in an outbreeding species full-sib family leads to a mixture of $lm \times ll$, $nn \times np$ and $hk \times hk$ segregation types. The two-way pseudo-testcross approach will lead to a large reduction in the sample size, as all heterozygous hk observations must be ignored. Therefore, with the application of SNPs in an outbreeding species full-sib family, an analysis that makes use of the multipoint approach is especially valuable.

8.6   Map estimation

In the two-way pseudo-testcross strategy, the more complicated analysis with several segregation types and two meioses is avoided by studying one of the two meioses at a time. The corresponding map estimation procedure is equivalent to that of a backcross population of an inbreeding species. The result will be two separate maps. Unless special restrictions are applied, these maps do not necessarily have the same order for the loci that appear on both maps. A locus segregating in both parents is called an *anchor* locus or a *bridge* locus. It is considered a reasonable assumption that the chromosomal order of loci is identical across individuals within a species. Chromosomal inversions and translocations are considered rare exceptions. Thus, normally one would expect identical orders of the two separate maps. Larger inversions and translocations in particular are likely to lead to irregular synapsis in early meiosis, resulting in chromatid breakage in the subsequent meiotic divisions and finally sterile gametes. Sometimes, a specially adapted linkage analysis may be possible (e.g. Farré *et al.*, 2011).

Combining the two separate parental maps into one, also called *map integration*, will allow us to estimate the order of loci segregating in only the first parent ($lm \times ll$) relative to those segregating in only the second parent ($nn \times np$). In

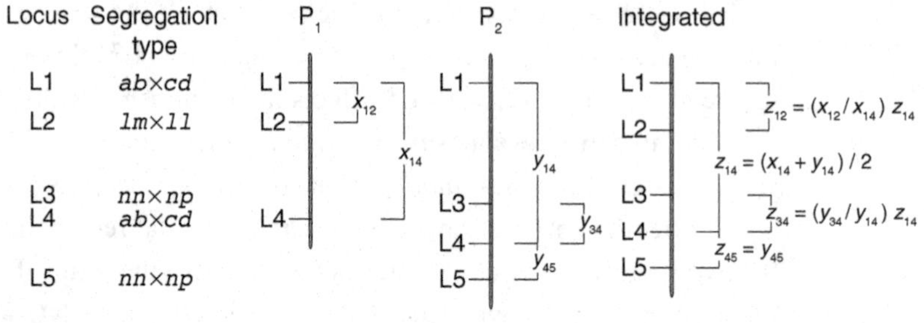

Locus	Segregation type	P$_1$	P$_2$	Integrated
L1	$ab \times cd$	L1	L1	L1
L2	$lm \times ll$	L2		L2
				$z_{12} = (x_{12}/x_{14})\, z_{14}$
		x_{14}	y_{14}	$z_{14} = (x_{14} + y_{14})\,/\,2$
L3	$nn \times np$		L3	L3
L4	$ab \times cd$	L4	L4	L4
			y_{34}	$z_{34} = (y_{34}/y_{14})\, z_{14}$
			L4	
L5	$nn \times np$		L5	$z_{45} = y_{45}$ L5

Fig. 8.4

Map integration of separate parental maps using simple geometry. **L1** and **L4** are anchor loci, which are present in both parental maps, P$_1$ and P$_2$. Map distances of the integrated map, z_{ij} between loci i and j, are calculated from the separate parental map distances, x_{ij} and y_{ij}. A segment between anchor loci obtains the average length (z_{14}). Non-anchor loci within the anchored segment (**L2**, **L3**) obtain linearly interpolated positions (z_{12}, z_{34}). A non-anchor locus (**L5**) outside the most distal anchor locus on the map (**L4**) retains the same distance to the distal anchor locus (z_{45}).

addition, applications such as QTL analysis of an outbreeding species full-sib family may require an integrated map. In order to accomplish the integration of the two separate maps, at least two anchor loci on the linkage group are necessary that lie a certain distance from each other. Having only a single anchor locus may allow the assignment of the separate parental linkage groups to the same linkage group; however, at least two distinct anchor loci are needed to obtain a proper orientation of the two groups with respect to each other. Having more than just two anchor loci will lead to better results.

The integration of the two pseudo-testcross maps can be done using simple geometry (Fig. 8.4). First, the distances between neighbouring anchor loci are calculated as the average of their parental map distances. Next, loci segregating in only one of the parents and positioned in between anchor loci are placed on the integrated map by linear interpolation according to the relative position between the flanking anchor markers on the relevant parental map. Finally, loci segregating in only one of the parents and positioned distal to the outermost anchor locus on the map are placed on the integrated map whilst retaining the distances to the anchor loci they had in the parental maps.

Another approach to obtaining the integrated map is to apply the linear regression method (see Section 5.3). There are actually two ways to apply this method, although one is preferred. The first way to apply it is to use the separate map distances as the dependent variable with the corresponding (squared) *LOD*s as weights. The disadvantage of this approach is the reduced utilization of the $hk \times hk$ type loci, which on average lose half their observations (i.e. the hk phenotypes). The second way is to use the estimates of the pairwise recombination frequencies assuming equal parental recombination probabilities as the dependent variable. In this approach, most of the $hk \times hk$ type locus observations are used wherever possible. There are two potential shortcomings to the linear regression method. They will arise in situations where there are larger differences in the separate parental recombination frequencies, which can be local or global. The first shortcoming is that such differences may lead to a poor fit of the data to the model and, as a result, negative map distances are estimated. Of course, negative distances cannot be interpreted other than that certain map segments should be inverted, but this, in turn, may be in conflict with the separate parental map orders. The second shortcoming is that, despite having non-conflicting orders in the separate parental maps, loci segregating in only one parent will have a better fit outside the anchored segment where it sits in the separate parental map. As a result, the integrated map order will be in conflict with the separate parental map order. This shortcoming may be prevented by placing the separate parental map orders as restrictions in the map integration algorithm.

Finally, the method of maximum likelihood can be used to obtain an integrated map. The method of Van Ooijen (2011) actually estimates the two separate parental maps simultaneously based on maximum likelihood, which is followed by the simple geometric map integration described above. The two separate parental maps are estimated using the multipoint estimation under a model of parent-specific recombination probabilities, as described in Section 8.5, and under the restriction of identical orders of the loci that are present in both maps. The multipoint estimation warrants that the maximum available information is used. Maximum likelihood warrants that no negative distances are estimated. Lastly, the used restriction of identical orders together with the geometric map integration warrants that there will be no conflict in the orders between the separate parental and integrated maps.

Exercises

Exercise 8.1.
A segregating outbreeding species full-sib family was coded using the knowledge of the alleles of the grandparents; therefore, the linkage phases could be included in the data. Later, for some reason, it was decided that grandparent one and grandparent two should be exchanged. What would have to be changed in the code of the segregation data? Explain your answer.

Exercise 8.2.
A segregating outbred full-sib family was coded, but later, for some reason, it was decided that parent one and parent two should be exchanged. What would

have to be changed in the code of the loci with segregation types $lm \times ll$, $nn \times np$, $ab \times cd$ and $hk \times hk$? Explain your answer.

Exercise 8.3.

(This exercise can be done with the evaluation version of JoinMap, which is freely available from http://www.joinmap.nl.)

An outbreeding species full-sib family was simulated with complete genotype data for five markers and a family size of 100; all markers were in coupling phase in both parents. The complete data are in the file *CP_abxcd.loc*. From this dataset, the separate parental datasets were derived; these are in the files *CP_abxcdP1.loc* and *CP_abxcdP2.loc*. In addition, a mixed configuration of segregation types was derived with the following types successively on the map: $ab \times cd$, $lm \times ll$, $nn \times np$, $ab \times cd$ and $hk \times hk$; these data are in the file *CP_mix.loc*. Load all datasets into a JoinMap project.

A. Verify that the separate parental data are indeed derived from the complete genotype data.
B. Verify that the mixed configuration data are indeed derived from the complete genotype data.
C. Determine the pairwise recombination frequency between markers M1 and M4 in the four datasets. How are the estimates related to each other?
D. Determine the pairwise recombination frequency between markers M1 and M2 in the four datasets. How are the estimates related to each other?
E. Determine the pairwise recombination frequency between markers M4 and M5 in the four datasets. How are the estimates related to each other?
F. Estimate the map of the mixed configuration with the regression method and with the maximum likelihood method and compare.

References

Burt, A., Bell, G. & Harvey, P. H. (1991). Sex differences in recombination. *Journal of Evolutionary Biology*, 4, 259–77.

Farré, A., Lacasa Benito, I., Cistué, L., De Jong, J. H., Romagosa, I. & Jansen, J. (2011). Linkage map construction involving a reciprocal translocation. *Theoretical and Applied Genetics*, 122, 1029–37.

Grattapaglia, D. & Sederoff, R. (1994). Genetic linkage maps of *Eucalyptus grandis* and *Eucalyptus urophylla* using a pseudo-testcross: mapping strategy and RAPD markers. *Genetics*, 137, 1121–37.

Jansen, J., De Jong, A. G. & Van Ooijen, J. W. (2001). Constructing dense genetic linkage maps. *Theoretical and Applied Genetics*, 102, 1113–22.

Lenormand, T. & Dutheil, J. (2005). Recombination difference between sexes: a role for haploid selection. *PLoS Biology*, 3, e63.

Maliepaard, C., Jansen, J. & Van Ooijen, J. W. (1997). Linkage analysis in a full-sib family of an outbreeding plant species: overview and consequences for applications. *Genetics Research*, 70, 237–50.

Ritter, E. & Salamini, F. (1996). The calculation of recombination frequencies in crosses of allogamous plant species with applications to linkage mapping. *Genetics Research*, 67, 55–65.

Van Ooijen, J. W. (2011). Multipoint maximum likelihood mapping in a full-sib family of an outbreeding species. *Genetics Research*, 93, 343–9.

Wu, R. L., Ma, C. X., Painter, I. & Zeng, Z. B. (2002a). Simultaneous maximum likelihood estimation of linkage and linkage phases in outcrossing species. *Theoretical Population Biology*, 61, 349–63.

Wu, R. L., Ma, C. X., Wu, S. S. & Zeng, Z. B. (2002b). Linkage mapping of sex-specific differences. *Genetics Research*, 79, 85–96.

9 Mapping in practice

Genetic linkage mapping is a very powerful tool, but it turns out to be quite sensitive to incomplete or erroneous information. In practice, it is often impossible to record data on all loci-individual combinations. Therefore, the mapping computations have to be done with some missing observations. It also turns out that a mapping experiment is prone to errors because of the huge number of observations. In this chapter, we address some of the common problems encountered in practice.

9.1 Introduction

The preceding chapters describe the theory underlying the construction of genetic linkage maps. With suitable computer software, map construction should be straightforward. We write *should*, because in practice this is not always the case. There are two reasons for this. As in statistical modelling, we view map construction as a method in which observations are fitted to a model. In practice, the fit between observations and model can be far from perfect. In this case, the observations may not behave according to the model or the model is an incorrect abstraction of the way the observations behave. In our specific case of genetic map construction, this means that a poor fit may either be caused by poor quality of the marker observations or by an imperfect model of the genetics, or both.

Investigating the reasons for a poor fit should be an essential part of the map construction process. Jansen *et al.* (2001a) embedded the actual construction of genetic linkage maps in a process of pre-mapping and post-mapping diagnostics. The mapping process is typically an iterative procedure, as depicted in Fig. 9.1. With pre-mapping diagnostics, the observations are inspected prior to the mapping computations; if necessary, they are corrected. After the mapping computations, the results are inspected using post-mapping diagnostics. If necessary, the computations are repeated, or the original data are corrected. Possibly, multiple iterations of this scheme will be necessary before reaching a reliable map as the final result. Because during the entire process you will make many decisions, it is sensible to keep a detailed log book.

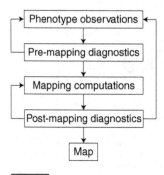

Schematic overview of the mapping process.

9.2   Model assumptions

In the theory description so far, we have not made the model assumptions very explicit. Usually the assumptions are indicated briefly as *Mendelian inheritance*. The actual meaning of this is that the inheritance is expected to behave according to the description of meiosis in Chapter 2. To be more specific, there are four major assumptions. The first concerns Mendel's Law of Segregation: half of the gametes of a heterozygous individual *Aa* are expected to carry allele *A* and the other half allele *a*. The second assumption involves Mendel's Law of Independent Assortment, which holds only for loci on different chromosomes: the distribution over the gametes of the alleles of locus **A** on one homologous pair is independent of that of the alleles of locus **B** on another homologous pair. The third assumption is about intrachromosomal recombination. In terms of probability theory, intrachromosomal recombination behaves as a Poisson process. Crossovers are assumed to take place between non-sister chromatids in the stage where the homologous chromosomes are duplicated into four chromatids. Crossovers occur independently, at random positions, at random frequency, and there is no preference as to which chromatids are involved. The preferential pairing of chromatids is called chromatid interference. Its absence ensures that the theoretical upper bound of the recombination frequency is $\frac{1}{2}$. The independence between crossovers is the same as absence of crossover interference. Absence of crossover interference is a requirement for setting up the likelihood function for three or more loci by simply multiplying probabilities of (no) recombination in segments between the loci (as is done in Chapter 5). The fourth and final assumption is that gametes (male and female) combine at random during the formation of zygotes (but of course depending on the type of breeding).

Although we do not yet know all the (molecular) details of recombination, our current knowledge of meiosis is sufficient to state, that under normal circumstances, the above assumptions are correct, with the exception that there usually is some crossover interference. Fortunately, Speed *et al.* (1992) have provided evidence for the robustness of the likelihood model for ordering markers with respect to crossover interference. Thus, different levels of crossover interference will not affect the estimated order, only the map distances. The differences between the Haldane and Kosambi mapping functions are only important for larger distances (>5 cM; see Fig. 5.1). Because most practical applications concern short distances only, this means that, for practical purposes, the level of crossover interference is not very important.

We write that the assumptions are correct 'under normal circumstances', because in some situations meiosis does not behave normally. This is due to chromosomal aberrations such as translocations, insertions and deletions. Such situations can lead to the formation of quadrivalents rather than bivalents during meiosis, which can result in chromatid breakage and subsequently reduced fertility of the gametes. Therefore, you may recognize these situations beforehand. In linkage analysis, they may result in spurious linkage and deviations from Mendel's Law of Segregation (e.g. Farré *et al.*, 2011).

9.3 Observation errors

The quality of the marker observations will depend primarily on the quality of the procedures with which mapping experiments are set up and carried out. A mapping experiment is a very extensive procedure, from selecting individuals that will become the parents of the experimental cross to making the computer data file with the marker phenotypes. The procedure is extensive in time as well as in number of handlings, and there may also be many people involved. All these aspects indicate that the final output of a mapping experiment, i.e. the phenotype observations, is prone to errors. Protocols for all parts of the experiment and high-quality administration will certainly contribute to the prevention of errors. Nevertheless, as in the proverb 'You cannot make an omelette without breaking eggs', making errors is inevitable in practice. Once errors are present in the final phenotype observations, they will affect the map construction and thus the quality of the resulting map. Therefore, detection of errors should be an integral part of the map construction procedure.

Errors can be random, such as accidental modifications of single observations. They can also be more systematic, such as when many observations of a (group of) individual(s) or marker(s) are recorded erroneously. Sometimes errors are so peculiar that the analysis cannot make sense of the data. An example is the situation where, accidentally, an entire batch of individuals in a biochemical analysis is treated differently from the rest of population. In all instances, it is helpful to have a good understanding of all handling procedures during the mapping experiment, so that it is possible to investigate and trace mistakes.

Normally, we assume that genetic markers relate exactly to one position on the entire genome of a species. A typical error occurs when a marker technique 'picks up signals' from two (or more) positions on the genome due to duplicated DNA segments. This situation occurs especially in allopolyploid species, where the genome has homeologous chromosomes with similar stretches of DNA. If this problem occurs in a single segregating population, the phenotypes will be translated erroneously into genotypes. For instance, if each of two loci segregate with the same two alleles, *A* and *a*, then all genotypes at the two loci except *AA,AA* and *aa,aa* will possess both alleles and will be scored as heterozygous if only a single segregating locus is assumed. Of course, the problem can be recognized beforehand if the parents are indistinct whilst at the same time there is segregation in the offspring, such as occurs with parental genotypes *AA,aa* and *aa,AA*. Therefore, it is important also to determine markers in the parents. When comparing markers across populations, it is always good to realize the possibility that they may reside in different positions in different populations due to variations in segregation patterns. Administrative errors may also cause the same problem.

What are the consequences of errors? This question is very difficult to answer in general. However, we can say something about random errors. If an error occurs at a certain pair of locus observations, then, if the underlying gamete is non-recombinant, we will observe it as recombinant, and vice versa. As most gametes are non-recombinant for a pair of loci (*Remark:* $r < \frac{1}{2}$), an error will most likely hit a non-recombinant and thus create an extra recombinant. Thus, random errors will increase the recombination frequency estimates. This effect is relatively larger for smaller distances. Consequently, the total map length will grow with the addition of more loci, even if the genome is already covered entirely with the current set of loci. Additionally, errors are likely to create local optima in the target function (i.e. the map evaluation criterion) that is optimized for finding the best order on the map. The more errors there are, the more local optima there will be in the target function and the greater the chances are that the order optimization will not arrive at the global optimum, whatever the optimization algorithm and whatever the target function (c.f. Lincoln & Lander, 1992).

The analysis should attempt to detect errors, but this is not always easy. By (statistically) testing whether the observations behave according to the assumptions, i.e. all aspects of Mendelian inheritance, it is sometimes possible to detect errors, or at least to point to highly unlikely observations. Ideally, you should inspect unlikely observations for errors throughout the entire trajectory from the start of the experiment to the data file, and correct them if found to be erroneous. Alternatively, you can remove the unlikely observations, which may be an acceptable option if you have an abundance of observations. Of course, preventing errors is the best option. Comparisons with independent repetitions of distinct parts or even an entire mapping experiment will give an approximate idea of the error rate in your current experimental setting.

9.4 Phenotype and genotype

The second aspect determining the quality of the marker observations concerns the power with which different genotypes can be distinguished. In fact, this is about the amount of information that is contained in marker observations. In Chapter 3, we noted that what we actually observe is the phenotype, and not the genotype. The phenotype is an observation of the underlying genotype. In many instances there is a 1:1 relationship, for instance in the backcross. However, there are also situations where there are uncertainties, for instance in the case of dominance. For a dominant observation, there are usually two possibilities for the underlying genotype. Sometimes special techniques can be used to resolve uncertainties; for instance, intensities or dosages of molecular marker signals can be used to separate phenotypes with a single copy from those with a double copy of a marker allele (e.g. Jansen *et al.*, 2001b). Nevertheless, even then, it may be impossible to classify intermediate intensities, so that a part of the observations remains in the dominant phenotype class. The uncertainties also concern the linkage phase. For instance, for an F_2 individual with heterozygous phenotypes at two loci, it is not known whether the loci are in coupling or repulsion phase. In coupling phase, there are no recombination events, whereas there are two in repulsion phase.

What is the actual problem with a small amount of information for a marker? The starting point of mapping is estimating the recombination frequency. In instances with high information content, the estimation comes down to determination, i.e. recombination events can simply be counted, but with low information content, the estimates will have a lower statistical confidence. An extreme example is that of dominantly observed markers in repulsion phase in the F_2 (see Chapter 3). The mapping of low-information markers will lead to map positions with a low confidence. A related aspect is that it is difficult to visualize all uncertainties that are present in a map.

Individual:	(a) 1	2	3	4	(b) 1	2	3	4
Locus **A**:	*AA*	*AA*	*Aa*	*Aa*	*AA*	*AA*	*UU*	*Aa*
Recombination:	x	x			x	x		
Locus **B**:	*Aa*	*UU*	*UU*	*Aa*	*Aa*	*Aa*	*Aa*	*Aa*
Recombination:			x	x			x	x
Locus **C**:	*Aa*	*Aa*	*AA*	*AA*	*Aa*	*UU*	*AA*	*AA*

Fig. 9.2

Inconsistency in recombination frequency estimates in a BC_1 population. Two possible configurations are shown, (a) and (b). Only the four individuals of the population that have a recombination event in the range **A–B–C** are shown. Genotype *AA* is like the backcross parent, genotype *Aa* is like the F_1 and genotype *UU* is unknown.

9.5   Missing observations

Missing observations are another aspect of the information contained in a marker or gene. If all loci have missing data for the same set of individuals, this simply reduces the population size (and the accuracy of estimates of all pairwise recombination frequencies). However, if the missing observations occur randomly across individuals and markers, this has three effects. First, estimates of recombination frequencies are based on different numbers of observations. Secondly, it may become impossible to locate recombination events. For instance, suppose, for the locus order **A–B–C**, a certain individual lacks the phenotype observation for **B**, whilst recombination is observed between **A** and **C**. It then remains unknown whether the recombination occurred in interval **A–B** or in interval **B–C**. Thirdly, the estimates of pairwise recombination frequency estimates will contain inconsistencies. Figure 9.2 illustrates this point for a backcross. It shows the genotypes of the only four individuals of the population that have a recombination event in the segment **A–B–C**. In configuration (a), we find the next recombinations: one between **A** and **B**, one between **B** and **C** and four between **A** and **C**. Disregarding the population size and the mapping function, the sum of the distances **A–B** and **B–C** is much smaller than the distance **A–C**. In configuration (b), the inconsistency is in the opposite direction: between all three pairs, we count two recombinations, and the sum of **A–B** and **B–C** is much larger than **A–C**. In theory, close inspection might resolve a few such problems, but in practice, with hundreds of loci and a hundred or more individuals, this will usually be considered too much work. As is the case with errors, inconsistencies like these are likely to create many local optima in the target function for order optimization. Larger numbers of missing observations will increase the chances that the order optimization will not arrive at the global optimum.

Finally, when a specific locus or specific individual has a considerable number of missing observations, the question arises as to the quality of the observations that could be made. Is it likely that their quality is poor? A poor-quality DNA sample of an individual may generate many missing observations and possibly also erroneous observations. Similarly, the poor quality of a certain locus, for example due to a minor mismatch of a DNA primer, may produce many missing and erroneous observations. In such instances, it is sensible to remove the marker or individual from the mapping process.

9.6 Pre-mapping diagnostics

Phenotype frequencies

Phenotype frequencies are the first to be inspected. There are two types: across loci and across individuals. For individuals, the distribution of phenotypes across loci allows the detection of individuals with many missing observations. It may be sensible to remove individuals with many missing observations. You should set a threshold for the acceptable level of missing observations, which should depend on the required quality of the data and the number of individuals. The distribution of the non-missing phenotypes will vary extensively across individuals. For instance, in an F_2, some individuals will be closer to the first parent, others to the second parent and yet others somewhere in between. However, an individual should never be identical to a parent; such chances are minute unless there are very few loci or you are dealing with a single chromosome. Thus, you should not test the distribution according to some expected segregation ratio, because there is no expected ratio. On the other hand, real outliers should be identified, as they could be the result of errors such as the accidental use of a parental seed in place of an offspring seed.

Segregation ratio

For loci, the distribution of phenotypes across individuals is usually called the segregation ratio. This distribution will reveal loci with many missing observations. Here, too, you should set a threshold for the acceptable level of missing observations. With regard to the non-missing observations, it is assumed that Mendel's

Law of Segregation applies. Normally, you may expect loci to segregate according to a certain ratio that depends on the type of population. This expected ratio is often called the Mendelian ratio. For instance, in an F_2, a ratio of 1:2:1 is expected for the genotypes *AA*, *Aa* and *aa*, respectively. A deviation from the expected ratio is called segregation distortion or skewed segregation. You can test whether the distribution is according to the expected ratio using Pearson's χ^2 test.

A (statistically significant) segregation distortion may indicate that some of the phenotypes are erroneous, for whatever reason. However, segregation distortion is also a phenomenon that is caused by selection, both natural and man-made. In plant science, for instance, so-called wide crosses are sometimes made, in which the parents are only remotely related or are even of different subspecies; in such crosses, segregation distortion is often observed. Selection can take place at the haploid level against a certain genotype of the gametes, and it can take place at the diploid level against a certain genotype of the zygotes or in later development stages. In populations involving multiple generations (e.g. RILs), selection can take place in successive generations, and even in different directions depending on the circumstances. The selection may concern the genotype at a single locus, but also at multiple loci simultaneously. The latter type can be responsible for spurious linkage between loci on non-homologous linkage groups. Due to linkage, loci near the selected locus will display the same level of segregation distortion, the level decreasing with distance. However, because distances are not known at this point in the analysis, this relationship will not be visible.

What must you do if you detect significant segregation distortion? Simply because of the potential problems associated with distorted segregation, you might consider it prudent to remove all loci displaying a certain amount of distortion. However, if the distortion is not due to error but due to natural selection, then by removing loci with distorted segregation you might be losing an entire linkage group or breaking a linkage group into subgroups. On the other hand, leaving distortedly segregating loci in may create difficulties in finding the proper linkage groups because of spurious linkage due to segregation distortion. Principal coordinate analysis may be an aid in such situations (Farré *et al.*, 2011). Another aspect is whether you have sufficient loci in the dataset to permit a stringent selection of loci whilst keeping sufficient coverage of the genome. Thus, in general, it will be a matter of trial and error. You may start with a stringent selection of loci, determine linkage groups and compute maps. Afterwards, you may add the loci, which you removed at first, and see what happens with the linkage group determination and the mapping. The alternative is to start with all markers, attempt to find the proper linkage groups, map them and verify whether the segregation distortion is visible on the linkage map as it should be: due to linkage, the pattern of the segregation distortion must display only gradual changes across the map.

Identical loci and individuals

Loci with identical phenotype observations in all individuals will map at the same position. Removing identical loci except one (to represent the identical loci in the mapping process) will reduce the computational workload and speed up the mapping process. The phenotypes must be identical for all individuals; also, if some observations are missing, they must be the same ones. Having the same missing observations is necessary, because if loci **A**, **B** and **C** have missing observations on different sets of individuals, they may have different recombination frequencies with other markers and should therefore be treated as different loci.

A typical error that is made, especially in populations derived from multiple generations of inbreeding, is the accidental duplication of an individual in a population, replacing another unique individual. The result is that each recombination event in that particular genotype will be duplicated, i.e. the map will be biased. Therefore, it is important to identify and remove individuals who have identical phenotype observations at all (or nearly all) loci. However, when studying a limited set of loci, or just a single linkage group, it is normal that some individuals will have identical phenotype observations. In this case, they should remain in the dataset.

Pairwise recombination frequencies

The starting point for ordering loci on the map is the set of pairwise, or two-point, recombination frequencies. One of the more common mistakes made when coding the observations on experimental populations derived from a cross between two fully homozygous parents is the accidental exchange of phenotype codes for a subset of loci, i.e. phenotypes that are actually like the first parent are coded as if being like the second parent and vice versa. If this mistake is made for all loci in the population, then there will be no problem in the mapping computations. Of course, you will make crucial errors after the mapping, when trying to select for certain genotypes. However, if the accidental exchange of phenotype codes is done for a subset only, then the recombination frequencies of these loci with correctly coded loci will attain the value 1 minus the correct value, which will be greater than $\frac{1}{2}$. If the linkage groups are determined using recombination frequencies, then the subset will become a group separate from their proper group. If, however, the test for independent segregation is used, then the subset will be grouped correctly, but it will be evident that the resulting map is going to be wrong. It depends on the mapping algorithm used as to what exactly happens in the map computations. Therefore, it is important that the estimates of pairwise recombination frequencies are inspected for this kind of mistake prior to the mapping, by looking for pairs with a recombination frequency greater than $\frac{1}{2}$. However, because random variation is also a cause of values greater than $\frac{1}{2}$, it will be better to set the detection threshold at a higher level, e.g. at 0.6. In populations derived from multiple generations of inbreeding, random variation has an even greater effect, so that in such cases the threshold may be set even higher.

9.7   Post-mapping diagnostics

After the map computations have been done, there are several inspections to make before the final result of the mapping process is obtained. With the inspections, you may detect problems with individuals, with loci and with the mapping algorithm. The diagnostics are targeted towards detecting erroneous observations and incorrect orderings. If necessary after detecting problems, you should make corrections to the data or, possibly, to the parameter settings of the mapping algorithm; afterwards, you can run the computations again to obtain an improved result.

Segregation distortion

Now that the loci are ordered according to their map position, you can study the segregation ratio in relation to the map position. Due to linkage, segregation distortion caused by selection should never display sudden changes along the map; if there are changes, they should always be gradual. Segregation distortion that is caused by errors in the phenotype observations of a locus will create a sudden change. However, because the errors also increase the recombination frequency estimates, the change will be accompanied by relatively large distances in the map. Therefore, the change along the map will be smoother than you might expect. Plotting phenotype frequencies or χ^2 test values against the map position might be the best way to detect any irregularities.

Number of recombinations per individual

After the loci are ordered on the map, it becomes possible to estimate for every individual the number of recombination events that have taken place, given the current map order. In population types where phenotypes are practically the same as genotypes, the estimation is merely counting. Otherwise, the estimation has to be done by maximum likelihood. Under normal circumstances, you will see a regular range in the number of recombination events. The range will depend on the length of the linkage group and on the number of meioses that have contributed to the segregation in the experimental population. For instance, in a backcross with a 100 cM map length, the range will be from zero to maybe four recombinations per individual, whilst in a RIL population, the same map length will show a range up to ten. Individuals that contain erroneous phenotype data will not fit well in the current map order and are likely to have a number of recombinations outside the range. If this is due to the poor quality of the DNA sample, for instance, then you should see this individual being outside the range for the other linkage groups as well.

When the range is longer than expected, this is an indication of an incorrect map order, i.e. the map optimization process finished at a local optimum. This extended range will occur together with a longer map than expected. This inspection requires that you know the approximate map length to expect, for example from other experiments.

Phenotype probabilities and graphical genotypes

Random errors in the phenotype observations are likely to stand out in the range of phenotypes of neighbouring loci on the map. Such phenotypes are usually called singletons. It will look as if the erroneous observation is flanked by recombination events, an unlikely situation. By calculating the probability for each observation, conditional on the observations at the flanking loci and conditional on the corresponding map distances, you can detect highly unlikely observations. You should not just modify unlikely observations without verification, because such situations do occur in reality. For instance, RIL and AIL families often have recombination events close to each other.

So-called graphical genotypes are a visual approach to detecting unlikely observations. For this, special software is freely available (Van Berloo, 2007), but similar images can also be produced with standard spreadsheet software using a feature called conditional formatting. The data matrix of phenotype observations is shown with the loci in the order of the map, whilst each data cell has a specific background colour that represents the phenotype. Figure 9.3 shows an example of graphical genotypes shown in greyscale. With contrasting colours, singletons will stand out even more visually. In addition to singletons, patterns that are more complex may also sometimes show up, which do not appear in the above probability calculations. For instance, individual number 43 in Fig. 9.3 has several switches of phenotype

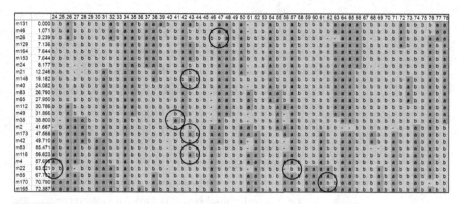

Example of graphical genotypes, shown in greyscale. Phenotype observations are shown as a data matrix with the individual number in the top row and the names of the loci and their map position in the first two columns. The circles indicate some examples of unlikely observations. Individual number 43 has several switches of phenotype along the map. Marker m22 is involved in two unlikely observations and also has several missing observations.

along the map, whilst marker m22 is involved in two unlikely observations and also has several missing observations.

Stress

Suppose we have three loci, **A**, **B** and **C**, in the order **A**–**B**–**C**, then we can predict according to our model the distance or recombination frequency between **A** and **C**, from the neighbouring segments **A**–**B** and **B**–**C**. We can compare this prediction with the estimate we have available for **A**–**C** directly. This difference is usually referred to as the nearest-neighbour stress, as it measures the discrepancy between the prediction from the underlying segments and the direct measurement. It can be measured in units of recombination frequency and in centiMorgans. If the observations behave according to our model assumptions, then the difference should be small. Because observations are discrete and finite, whereas the probability model of recombination is continuous, there will usually be a minor difference. Now suppose there are errors in the observations of the middle locus, **B**, then these will inflate the estimates of the recombination frequencies with **A** and **C**, but they will not affect that of **A** with **C**. Obviously, this will generate nearest-neighbour stress for locus **B**. Errors in locus **A** or **C** will affect all three recombination frequency estimates and thus will not generate much stress. Any significant level of crossover interference might also be the cause of stress, but because for instance the difference between the Haldane and Kosambi mapping functions is only important when dealing with large distances, interference is considered not to be a very likely cause of stress.

How much stress can be tolerated? It is difficult to answer this question, as there is no statistical test. If a particular locus is troubled by errors, the stress value of the locus will hopefully stand out compared with other loci. In gauging the importance of stress, the density of the map is relevant. The same value of stress (e.g. 2 cM) has a much greater impact in a very dense map than in a sparse map (with markers at about every 10 cM).

Model fit

When fitting data to a model, it is pleasant to have a measure of quality of the fit. In the case of a poor fit, indicators to the cause(s) are very useful. When using the maximum likelihood mapping algorithm, the log-likelihood value can be used to compare different map orders. However, the log-likelihood value itself depends on the number of observations (loci and individuals) and is not a suitable measure for the quality of the fit. In Chapter 6, we described the use of the χ^2 goodness of fit in the regression-mapping algorithm. Because the measure behaves approximately as a χ^2 variable, it can be used as a measure of the fit of the data to the model. Interestingly, the contributions of the separate loci to the statistic can be determined, so that in cases of a poor fit the responsible loci may be detected (unless all loci are equally responsible).

In the regression-mapping algorithm, a filter must be used to select the pairs of loci for use in the regression. This is necessary because recombination frequencies close to and greater than $\frac{1}{2}$ are unsuitable for translating into centiMorgan distances, which are the variables on which the regression is based. If the filter is set to a stringent value, the algorithm will 'focus' on the more local distances. A problem that may occur in such a setting is that loci that do not fit well at their proper position can become located outside the region, of which pairwise distances with this locus are included. Due to the stringent filter, the goodness of fit calculated for such an order does not include the distances that would indicate the erroneous positioning of the locus. As a result, the optimization will attain a better goodness of fit at this position than at the proper position. By calculating a fit measure based on the flanking loci (in the final order) and ignoring the filter, thus including all pairwise distances, an indication can be obtained of how well individual loci fit at their estimated map position. A simple statistic subtracting the pairwise recombination

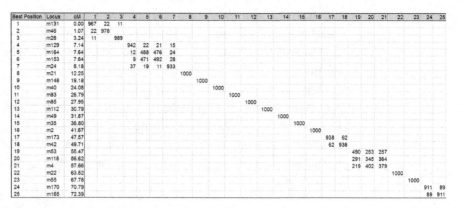

Fig. 9.4

Example of plausible map positions. In the set of 1000 Metropolis samples drawn, the positions of all loci in each map are recorded. These positions are plotted against the position of the best map found.

frequencies from the map-based recombination frequencies of a locus with its flanking loci can be called the nearest-neighbour fit. Such a statistic can also be based on centiMorgans. A poor fit not only indicates that a locus is placed at the wrong position but also that it does not fit very well at its proper position. Whether the latter is due to the locus itself or its neighbours remains the question. Of course, placing a locus at the wrong position may also affect the correct placement of other loci.

Plausible map orders

Normally, genetic linkage maps are published without any indication of their accuracy. The static nature of a map may give a false impression of its reliability. The accuracy depends on the population type, the number of individuals and the quality of the observations. Jansen *et al.* (2001a) introduced the concept of 'plausible maps' as a means of visualizing at least a part of the (in)accuracy. Plausible maps are map orders that, although not optimal, could also have been obtained with the current set of data. Starting from the optimal map, new map orders are sampled using a Metropolis algorithm. For example, for every 1000th sampled map, the positions of all loci are recorded, and after 1000 of such recorded samples, these plausible positions of all loci can be plotted against the best position. A normal result should show a pattern of positions around the main diagonal, where loci primarily exchange positions with their nearby neighbours, e.g. Fig. 9.4. A wider range around the diagonal indicates greater uncertainty. If the pattern is very irregular, then this is an indication that the order optimization algorithm may have ended up in a suboptimal map order.

References

Farré, A., Lacasa Benito, I., Cistué, L., De Jong, J. H., Romagosa, I. & Jansen, J. (2011). Linkage map construction involving a reciprocal translocation. *Theoretical and Applied Genetics*, 122, 1029–37.

Jansen, J., De Jong, A.G. & Van Ooijen, J.W. (2001a). Constructing dense genetic linkage maps. *Theoretical and Applied Genetics*, 102, 1113–22.

Jansen, R. C., Geerlings, H., Van Oeveren, J. C. & Van Schaik, R. C. (2001b). A comment on codominant scoring of AFLP markers. *Genetics*, 158, 925–6.

Lincoln, S. E. & Lander, E. (1992). Systematic detection of errors in genetic linkage data. *Genomics*, 14, 604–10.

Speed, T. P., McPeek, M. S. & Evans, S. N. (1992). Robustness of the no-interference model for ordering genetic markers. *Proceedings of the National Academy of Sciences USA*, 89, 3103–6.

Van Berloo, R. (2007). GGT 2.0: versatile software for the visualization of genetic data. *Journal of Heredity*, 99, 232–6.

Answers to exercises

Please note that the calculations of some of the exercises in this book are demonstrated with MS-Excel spreadsheets (.xls). These are included in the set of files that are available for downloading at the Cambridge University Press web site at http://www.cambridge.org/9781107013216.

Exercise 3.1.

First determine, for each combination of markers, which genotype classes are recombinant and which non-recombinant. Here, combinations of *AA* with *Aa* are recombinant, whilst combinations of *AA* with *AA* and *Aa* with *Aa* are non-recombinant. Next, determine the total number of recombinant individuals and divide this by the total number of individuals with known genotype:

K–L: $\hat{r} = (4 + 5 + 8 + 5)/100 = 0.22$
K–M: $\hat{r} = (0 + 5 + 8 + 1)/100 = 0.14$
L–M: $\hat{r} = (0 + 4 + 5 + 1)/100 = 0.10$

Exercise 3.2.

First determine, for each combination of loci, which genotype classes are recombinant and which non-recombinant. Here, combinations of *a* with *h* are recombinant, whilst combinations of *a* with *a* and *h* with *h* are non-recombinant. Any combination with an unknown observation, '–', must be subtracted from the 164 individuals to determine the total number of individuals with known genotype, as these combinations cannot be classified as recombinant nor as non-recombinant. Next, determine the total number of recombinant individuals and divide this by the total number of individuals with known genotype. The easiest way to do this is tallying the individuals in a 3×3 contingency table. If you have access to the data in a computer spreadsheet, you can achieve this easily with pivot tables.

	B		
A	*a*	*h*	–
a	64	13	1
h	6	53	5
–	6	10	6

$\hat{r} = (13 + 6)/(64 + 13 + 6 + 53) = 0.140$

	C		
A	*a*	*h*	–
a	54	20	4
h	15	47	2
–	6	9	7

$\hat{r} = (20 + 15)/(54 + 20 + 15 + 47) = 0.257$

	D		
A	*a*	*h*	–
a	62	15	1
h	11	51	2
–	5	7	10

$\hat{r} = (11 + 15)/(62 + 11 + 15 + 51) = 0.187$

	C		
B	*a*	*h*	–
a	59	14	3
h	12	60	4
–	4	2	6

$\hat{r} = (14 + 12)/(59 + 14 + 12 + 60) = 0.179$

	D		
B	*a*	*h*	–
a	70	6	0
h	4	65	7
–	4	2	6

$\hat{r} = (4 + 6)/(70 + 4 + 6 + 65) = 0.069$

	D		
C	*a*	*h*	*–*
a	67	7	1
h	8	62	6
–	3	4	6

$\hat{r} = (8 + 7)/(67 + 8 + 7 + 62) = 0.104$

Exercise 3.3.

A. $p_{re} = r^2/(s^2 + r^2) = 0.25^2/(0.25^2 + 0.75^2) = 0.1$

B. $E(R_{Aa,Bb}) = 2p_{re}n_{Aa,Bb} = 2 \times 0.1 \times 44 = 8.8$

C. $R' = 0 \times 18 + 1 \times 4 + 2 \times 1 + 1 \times 5 + 1 \times 1 + 2 \times 0 + 1 \times 4$
    $+ 0 \times 23 = 16$

D. $E(R) = R' + E(R_{Aa,Bb}) = 16 + 8.8 = 24.8$

E. $E(r) = E(R)/(2 \times 100) = 0.124$

F. $p_{re} = 0.0196, E(R_{Aa,Bb}) = 1.7286, R' = 16, E(R) = 17.7286, E(r) = 0.0886$

G. $p_{re} = 0.0094, E(R_{Aa,Bb}) = 0.8274, R' = 16, E(R) = 16.8274, E(r) = 0.0841$

In the two next iterations, $E(r) = 0.0837$ and 0.0836, respectively, whilst in subsequent iterations, $E(r)$ does not change any more.

Exercise 3.4.

A, B.

Locus 1	Locus 2	$\hat{r}$	LOD
m1	m2	0.1575	13.61
m1	m3	0.2460	7.33
m1	m4	0.3563	2.09
m2	m3	0.1555	15.05
m2	m4	0.2665	5.94
m3	m4	0.1268	18.68

C.

	m2			m3			m4	
m1	*AA*	*a–*	**m1**	*AA*	*a–*	**m1**	*AA*	*a–*
AA	18	11	*AA*	18	11	*AA*	15	14
a–	6	65	*a–*	10	61	*a–*	15	56

m2	**m3**		m2	**m4**		m3	**m4**	
	AA	*a−*		*AA*	*a−*		*AA*	*a−*
AA	19	5	*AA*	16	8	*AA*	23	5
a−	9	67	*a−*	14	62	*a−*	7	65

D.

Locus 1	Locus 2	$\hat{r}$	LOD
m1	m2	0.1824	6.48
m1	m3	0.2222	4.94
m1	m4	0.3150	2.01
m2	m3	0.1488	8.26
m2	m4	0.2412	3.99
m3	m4	0.1201	10.95

The recombination frequencies are somewhat different, and the *LODs* are much smaller.

E.

m1	**m2**		m1	**m3**		m1	**m4**	
	A−	*aa*		*AA*	*a−*		*A−*	*aa*
AA	29	0	*AA*	18	11	*AA*	26	3
a−	47	24	*a−*	10	61	*a−*	49	22

m2	**m3**		m2	**m4**		m3	**m4**	
	AA	*a−*		*A−*	*aa*		*A−*	*aa*
A−	28	48	*A−*	63	13	*AA*	28	0
aa	0	24	*aa*	12	12	*a−*	47	25

F.

Locus 1	Locus 2	$\hat{r}$	*LOD*
m1	m2	0.0000	4.22
m1	m3	0.2222	4.94
m1	m4	0.3162	1.14
m2	m3	0.0000	4.04
m2	m4	0.2965	2.09
m3	m4	0.0000	4.22

The recombination frequencies in the repulsion phase situations are often zero, and their *LOD*s are even smaller than in (D).

Exercise 4.1.

A. $\hat{r} = R/N = (17 + 10)/100 = 0.27$

B. $LOD = S \log_{10}(s) + R \log_{10}(r) + N \log_{10}(2)$
$= 73 \log_{10}(0.73) + 27 \log_{10}(0.27) + 100 \log_{10}(2)$
$= 4.77$

C. Complete the contingency table with the sums of the columns and rows:

	Locus B		
Locus A	*Bb*	*bb*	**Total**
Aa	32	17	49
aa	10	41	51
Total	42	58	100

Now you can see that the row probabilities are 49/100 and 51/100, respectively, and the column probabilities are 42/100 and 58/100, respectively. Thus, the expected cell (genotype combination) counts are:

	Locus B	
Locus A	*Bb*	*bb*
Aa	$49 \times 42/100 = 20.58$	$49 \times 58/100 = 28.42$
aa	$51 \times 42/100 = 21.42$	$51 \times 58/100 = 29.58$

From the observed and expected cell counts, you can calculate the X^2 and G^2 statistics:

$$X^2 = (32 - 20.58)^2/20.58 + (17 - 28.42)^2/28.42 +$$
$$(10 - 21.42)^2/21.42 + (41 - 29.58)^2/29.58 = 21.4$$

D. $G^2 = 2(32 \ln(32/20.58) + 17 \ln(17/28.42) +$
$$10 \ln(10/21.42) + 41 \ln(41/29.58)) = 22.3$$

E. When you use a (not very stringent) threshold of 3.0 for the *LOD*, then the loci are concluded to be linked. When using the X^2 and G^2 statistics for testing the independence between the loci, the loci are concluded not to segregate independently ($P(\chi^2_1 > 15.137) = 0.0001$), which is assumed to be caused by linkage.

Exercise 4.2.

A. $\hat{r} = 0.5$

B. $LOD = 0.0$

C. $X^2 = 11.11$

D. $G^2 = 17.0$ (Note: $(\lim x{\downarrow}0)\, x \log(x/N) = 0$)

E. This is a strange pair of markers, with contradictory linkage tests. Further investigation is needed into the cause of these results before continuing with the mapping of the markers.

Exercise 4.3.

A. $LOD = 5.60$

B. $X^2 = 26.21$

C. $G^2 = 25.67$

D. Degrees of freedom $= (3 - 1)(2 - 1) = 2$

E. Compare with the χ^2 distribution with 2 degrees of freedom. The P value of both statistics is less than 0.00001, so you can conclude that the markers are linked.

Exercise 5.1.

First, translate the recombination frequencies into map distances. If you use Haldane's mapping functions, you obtain:

$$d_{AB} = -50 \ln(1 - 2 \times 0.08) = 8.72\,\text{cM}, \quad d_{BC} = 6.39 \quad \text{and} \quad d_{AC} = 13.72\,.$$

Place these three values, or better, references to the cells where they are calculated, into a column in the spreadsheet (the following procedures are identical in MS Excel and OpenOffice Calc). Create next to this two columns with the indicator variables, 0 or 1, for the two distances to be estimated depending on the order. You should have the following 3×3 ranges:

Order A–B–C:			Order B–A–C:			Order A–C–B:		
8.72	1	0	8.72	1	0	8.72	1	1
6.39	0	1	6.39	1	1	6.39	0	1
13.72	1	1	13.72	0	1	13.72	1	0

Now you apply the LINEST() function. This is a so-called array function that requires a range of cells for its output. Select an empty range of five rows and two columns and enter: '=LINEST([known_y's], [known_x's],false,true) [control-shift-enter]', in which the [known_y's] must point to the range with the three map distances and the [known_x's] to the range with the indicator variables, e.g. '=LINEST(B9:B11,C9:D11,FALSE,TRUE)'. Finishing with [control–shift–enter] applies the formula to the selected range. In the top row, the two estimated map distances will be shown in reverse order:

Order $\mathbf{A-B-C}$: $\hat{\delta}_{AB} = 8.26$ and $\hat{\delta}_{BC} = 5.93$

Order $\mathbf{B-A-C}$: $\hat{\delta}_{AB} = 3.37$ and $\hat{\delta}_{AC} = 8.37$

Order $\mathbf{A-C-B}$: $\hat{\delta}_{AC} = 9.92$ and $\hat{\delta}_{BC} = 2.59$

Exercise 5.2.

Translate the recombination frequencies into map distances. If you use Haldane's mapping functions, you obtain:

Pair	$\hat{r}$	cM
A–B	0.140	16.43
A–C	0.257	36.08
A–D	0.187	23.42
B–C	0.179	22.16
B–D	0.069	7.43
C–D	0.104	11.66

Place references to these six distance values in a column in a spreadsheet. Create next to it the range of six rows and three columns with indicator variables, depending on the map order:

Order A–B–D–C:				Order A–D–B–C:			
16.43	1	0	0	16.43	1	1	0
36.08	1	1	1	36.08	1	1	1
23.42	1	1	0	23.42	1	0	0
22.16	0	1	1	22.16	0	0	1
7.43	0	1	0	7.43	0	1	0
11.66	0	0	1	11.66	0	1	1

After applying the LINEST() function, the three estimated map distances will be shown in reverse order:

Order $\mathbf{A-B-D-C}$: $\hat{\delta}_{AB} = 15.69, \hat{\delta}_{BD} = 8.09$ and $\hat{\delta}_{CD} = 12.68$

Order $\mathbf{A-D-B-C}$: $\hat{\delta}_{AD} = 20.06, \hat{\delta}_{BD} = -0.66$ and $\hat{\delta}_{BC} = 17.05$

Exercise 5.3.

Create the vector RFs with the six recombination frequencies:

```
> RFs <- c(0.140, 0.257, 0.187, 0.179, 0.069, 0.104)
```

Apply the defined function haldane() to this vector to produce the vector PDs of distances:

```
> PDs <- haldane(RFs)
> PDs
[1] 16.42520 36.07733 23.42025 22.15835 7.42500 11.65969
```

Create the three design vectors X1, X2 and X3 that apply to the order **B–C–A–D**:

```
> X1 <- c(1, 1, 1, 0, 0, 0)
> X2 <- c(0, 1, 1, 1, 1, 0)
> X3 <- c(0, 1, 0, 1, 0, 1)
```

Create the vector with the six *LODs* and the vector LODs2 with their squares:

```
> LODs <- c(17.05, 7.26, 12.75, 14.03, 27.85, 22.45)
> LODs2 <- LODs^2
> LODs^2
[1] 290.7025 52.7076 162.5625 196.8409 775.6225 504.0025
```

Calculate the weighted least squares model:

```
> WLS <- lm(PDs ~ X1 + X2 + X3 - 1, weights = LODs2)
```

With the command summary(WLS), you can find the three estimated distances:

$$\hat{\delta}_{AB} = 16.1084, \quad \hat{\delta}_{BD} = 7.8027 \quad \text{and} \quad \hat{\delta}_{DC} = 12.3993 \,.$$

For the order **A–D–B–C**, the three design vectors X1, X2 and X3 are (only X2 is different):

```
> X1 <- c(1, 1, 1, 0, 0, 0)
> X2 <- c(1, 1, 0, 0, 1, 1)
> X3 <- c(0, 1, 0, 1, 0, 1)
```

Recalculating the weighted least squares model produces the next estimated distances:

$$\hat{\delta}_{AD} = 17.114, \quad \hat{\delta}_{DB} = 3.410 \quad \text{and} \quad \hat{\delta}_{BC} = 12.393 \,.$$

Exercise 5.4.

Because there are no missing observations, you can calculate the log-likelihood based on the pairwise recombination frequencies:

$$\hat{r}_{DE} = 29/100 = 0.29, \ \hat{r}_{DF} = 16/100 = 0.16 \text{ and}$$
$$\hat{r}_{EF} = 17/100 = 0.17.$$

For the log-likelihood of the order **D–E–F**, you use the recombination of the segments **D–E** and **E–F**:

$$\ln(L) = 71 \ln(0.71) + 29 \ln(0.29) + 84 \ln(0.84) + 16 \ln(0.16) - 100 \ln(2)$$
$$= -175.12.$$

Exercise 6.1.

A. Order **A–B–C**: $SARF = \hat{r}_{AB} + \hat{r}_{BC} = 0.08 + 0.06 = 0.14$
   Order **B–A–C**: $SARF = \hat{r}_{AB} + \hat{r}_{AC} = 0.08 + 0.12 = 0.20$
   Order **A–C–B**: $SARF = \hat{r}_{AC} + \hat{r}_{BC} = 0.12 + 0.06 = 0.18$
B. The residual sum of squares is given in the cell in column 2, row 5 of the LINEST() results:
   Order **A–B–C**: $SS_R = 0.64$
   Order **B–A–C**: $SS_R = 85.84$
   Order **A–C–B**: $SS_R = 43.29$
C. The G^2 compares the two-point recombination frequency estimates ($\hat{r}_{tp}$) with the corresponding map-derived recombination frequencies over all pairs. To obtain the map-derived recombination frequencies ($\hat{r}_{mp}$), you apply the inverse mapping function (of Haldane) on the distance estimates ($\hat{\delta}_{mp}$).

Segment	$\hat{r}_{tp}$	$\hat{\delta}_{mp}$	$\hat{r}_{mp}$	$\hat{\delta}_{mp}$	$\hat{r}_{mp}$	$\hat{\delta}_{mp}$	$\hat{r}_{mp}$
		A–B–C		B–A–C		A–C–B	
A–B	0.08	8.26	0.076	3.37	0.033	9.92 + 2.59	0.111
B–C	0.06	5.93	0.056	3.37 + 8.37	0.105	2.59	0.025
A–C	0.12	8.26 + 5.93	0.123	8.37	0.077	9.92	0.090

   Order **A–B–C**: $G^2 = 0.064$
   Order **B–A–C**: $G^2 = 9.841$
   Order **A–C–B**: $G^2 = 5.610$
D. The order **A–B–C** is the best according to each of the three criteria above.

Exercise 6.2.

A. Order **A–B–D–C**: $SARF = \hat{r}_{AB} + \hat{r}_{BD} + \hat{r}_{CD}$
   $$= 0.140 + 0.069 + 0.104 = 0.313$$

Order **A–D–B–C**: $SARF = \hat{r}_{AD} + \hat{r}_{BD} + \hat{r}_{BC}$
$$= 0.187 + 0.069 + 0.179 = 0.435$$

B.   Order **A–B–D–C**: $SS_R = 4.23$
Order **A–D–B–C**: $SS_R = 134.12$

C.   In these data, the total number of gametes (N_p) varies over the pairs, so this needs to be taken into account. Below is a table with the parameters relevant for the calculations.

Segment	$\hat{r}_{tp}$	R_p	N_p	A–B–D–C $\hat{\delta}_{mp}$	A–B–D–C $\hat{r}_{mp}$	A–D–B–C $\hat{\delta}_{mp}$	A–D–B–C $\hat{r}_{mp}$
A–B	0.140	19	136	15.69	0.135	20.06 − 0.66	0.161
A–C	0.257	35	136	15.69 + 8.09 + 12.68	0.259	20.06 − 0.66 + 17.05	0.259
A–D	0.187	26	139	15.69 + 8.09	0.189	20.06	0.165
B–C	0.179	26	145	8.09 + 12.68	0.170	17.05	0.144
B–D	0.069	10	145	8.09	0.075	|−0.66|	0.007
C–D	0.104	15	144	12.68	0.112	−0.66 + 17.05	0.140

The second order has a negative distance estimate, which is difficult to work with. For the calculation of G^2, you take the absolute value of the map distance to get $\hat{r}_{mp}$.

Order **A–B–D–C**: $G^2 = 0.28$
Order **A–D–B–C**: $G^2 = 33.43$

D.   The order **A–B–D–C** is the best according to each of the three criteria above. The fact that in the other order a negative distance is estimated makes the order of suspicious quality.

Exercise 6.3.

A. $\hat{r}_{DE} = 29/100 = 0.29$
$\hat{r}_{DF} = 16/100 = 0.16$
$\hat{r}_{EF} = 17/100 = 0.17$

B.   Order **D–E–F**: $\hat{r}_{DE} + \hat{r}_{EF} = 0.46$;   difference with $\hat{r}_{DF}$ is 0.30
Order **D–F–E**: $\hat{r}_{DF} + \hat{r}_{EF} = 0.33$;   difference with $\hat{r}_{DE}$ is 0.04
Order **E–D–F**: $\hat{r}_{DE} + \hat{r}_{DF} = 0.45$;   difference with $\hat{r}_{EF}$ is 0.28
The order **D–F–E** appears to be the best map order because the sum of $\hat{r}$ values has the smallest discrepancy with the direct estimate.

C.   If you use the assumption of no crossover interference you will get:
Order **D–E–F**: $\hat{r}_{DE} + \hat{r}_{EF} - 2\hat{r}_{DE}\hat{r}_{EF} = 0.36$;   difference with $\hat{r}_{DF}$ is 0.20
Order **D–F–E**: $\hat{r}_{DF} + \hat{r}_{EF} - 2\hat{r}_{DF}\hat{r}_{EF} = 0.28$;   difference with $\hat{r}_{DE}$ is 0.01

Order **E–D–F**: $\hat{r}_{DE} + \hat{r}_{DF} - 2\,\hat{r}_{DE}\,\hat{r}_{DF} = 0.36$;   difference with $\hat{r}_{EF}$ is 0.19
The order **D–F–E** appears to be the best map order because of the smallest discrepancy with the direct estimate.

If you use the assumption of crossover interference according to Kosambi, you will get:

Order **D–E–F**: $(\hat{r}_{DE} + \hat{r}_{EF})/(1 + 4\,\hat{r}_{DE}\,\hat{r}_{EF}) = 0.38$;
  difference with $\hat{r}_{DF}$ is 0.22

Order **D–F–E**: $(\hat{r}_{DF} + \hat{r}_{EF})/(1 + 4\,\hat{r}_{DF}\,\hat{r}_{EF}) = 0.30$;
  difference with $\hat{r}_{DE}$ is 0.01

Order **E–D–F**: $(\hat{r}_{DE} + \hat{r}_{DF})/(1 + 4\,\hat{r}_{DE}\,\hat{r}_{DF}) = 0.38$;
  difference with $\hat{r}_{EF}$ is 0.21

The order **D–F–E** appears to be the best map order because of the smallest discrepancy with the direct estimate.

D. Order **D–E–F**: $\ln(L) = -175.12$
Order **D–F–E**: $\ln(L) = -158.87$
Order **E–D–F**: $\ln(L) = -173.50$
The order **D–F–E** appears to be the best map order because of the highest (least negative) log-likelihood.

E. Order **D–E–F**:   46
Order **D–F–E**:   33
Order **E–D–F**:   45
The order **D–F–E** appears to be the best map order because it has the smallest number of recombination events.

F. $LOD_{DE} = 3.95,$ $LOD_{DF} = 11.01,$ $LOD_{EF} = 10.30$
Order **D–E–F**: $SALOD = 14.25$
Order **D–F–E**: $SALOD = 21.31$
Order **E–D–F**: $SALOD = 14.96$
The order **D–F–E** appears to be the best map order because of the highest sum of LODs.

Exercise 6.4.

A. $\hat{r}_{XY} = 0.18$
$\hat{r}_{YZ} = 0.18$
$\hat{r}_{XZ} = 0.08$

B. Because there are no missing observations, you can calculate the log-likelihood based on the pairwise recombination frequencies:
$\ln(L_{X-Y-Z}) = 18\ln(0.18) + 82\ln(0.82) +$
  $18\ln(0.18) + 82\ln(0.82) - 100\ln(2) = -163.59$
$\ln(L_{X-Z-Y}) = \ 8\ln(0.08) + 92\ln(0.92) +$
  $18\ln(0.18) + 82\ln(0.82) - 100\ln(2) = -144.33$

$$\ln(L_{Y-X-Z}) = 18 \ln(0.18) + 82 \ln(0.82) +$$
$$8 \ln(0.08) + 92 \ln(0.92) - 100 \ln(2) = -144.33$$

X–Z–Y and **Y–X–Z** are equally good.

C. The number of recombinations between adjacent markers is:

      X–Y–Z: 36

      X–Z–Y: 26

      Y–X–Z: 26

      X–Z–Y and **Y–X–Z** are equally good.

D. $LOD_{XY} = 9.63$

   $LOD_{YZ} = 9.63$

   $LOD_{XZ} = 18.00$

   $SALOD_{X-Y-Z} = 19.26$

   $SALOD_{X-Z-Y} = 27.63$

   $SALOD_{Y-X-Z} = 27.63$

   X–Z–Y and **Y–X–Z** are equally good.

E. If you use the (unweighted) linear model in R or LINEST() in MS Excel or OO-Calc and as mapping function Haldane's:

Segment	$\hat{r}_{tp}$	X–Y–Z		X–Z–Y		Y–X–Z	
		$\hat{\delta}_{mp}$	$\hat{r}_{mp}$	$\hat{\delta}_{mp}$	$\hat{r}_{mp}$	$\hat{\delta}_{mp}$	$\hat{r}_{mp}$
X–Y	0.18	10.34	0.093	25.22	0.198	19.41	0.161
Y–Z	0.18	10.34	0.093	19.41	0.161	25.22	0.198
X–Z	0.08	20.68	0.169	5.81	0.054	5.81	0.054

      X–Y–Z: $G^2 = 21.1$

      X–Z–Y: $G^2 = 1.5$

      Y–X–Z: $G^2 = 1.5$

      X–Z–Y and **Y–X–Z** are equally good.

If you use the weighted linear model in R:

Segment	$\hat{r}_{tp}$	X–Y–Z		X–Z–Y		Y–X–Z	
		$\hat{\delta}_{mp}$	$\hat{r}_{mp}$	$\hat{\delta}_{mp}$	$\hat{r}_{mp}$	$\hat{\delta}_{mp}$	$\hat{r}_{mp}$
X–Y	0.18	6.61	0.062	26.13	0.203	18.50	0.155
Y–Z	0.18	6.61	0.062	18.50	0.155	26.13	0.203
X–Z	0.08	13.22	0.116	7.63	0.071	7.63	0.071

$$\mathbf{X–Y–Z}: \quad G^2 = 34.1$$
$$\mathbf{X–Z–Y}: \quad G^2 = 0.9$$
$$\mathbf{Y–X–Z}: \quad G^2 = 0.9$$

X–Z–Y and **Y–X–Z** are equally good.

Exercise 8.1.

The phase type codes of all loci should be changed; if the first digit is 0 it must become 1 and vice versa:

Change 00 into 10.
Change 10 into 00.
Change 01 into 11.
Change 11 into 01.

However, this is not really necessary because all recombination frequencies will be estimated the same.

Exercise 8.2.

The alleles present in the first parent now become those of the second parent and vice versa. You have to make use of the fixed set of segregation type codes. This means for the loci with segregation type:

ab×*cd*:
Do not change the segregation type.
In the genotypes, exchange the *a* with the *c* and the *b* with the *d* allele, thus:
Change *ac* into *ca*, which is the same as *ac*, so effectively there is no change.
Change *ad* into *cb*, which is the same as *bc*.
Change *bc* into *da*, which is the same as *ad*.
Change *bd* into *db*, which is the same as *bd*, so effectively there is no change.

hk×*hk*:
Because this segregation type is symmetric, it need not be changed and neither must the genotype codes be changed.

lm×*ll*:
Change the segregation type to *nn*×*np*.
Exchange the genotypes *ll* with *nn* and *lm* with *np*.

nn×np:
Change the segregation type to *lm×ll*.
Exchange the genotypes *nn* with *ll* and *np* with *lm*.

q0×q0:
Because this segregation type is symmetric, it need not be changed and neither must the genotype codes be changed,

rs×r0:
Change the segregation type to *t0×tu*.
Exchange the *r* with the *t*.
Exchange the *s* with the *u*.

t0×tu:
Change the segregation type to *rs×r0*.
Exchange the *t* with the *r*.
Exchange the *u* with the *s*.

Exercise 8.3.
A.   The translations are:

Complete data	To P1 data	To P2 data
ac:	*ll*	*nn*
ad:	*ll*	*np*
bc:	*lm*	*nn*
bd:	*lm*	*np*

B.   The translations are:

Complete data	To *lm×ll*	To *nn×np*	To *hk×hk*
ac:	*ll*	*nn*	*hh*
ad:	*ll*	*np*	*hk*
bc:	*lm*	*nn*	*hk*
bd:	*lm*	*np*	*kk*

C.   The pairwise recombination frequency estimates between markers M1 and M4 in the four datasets are:

Complete data: 0.1400
P1:   0.0700
P2:   0.2100
Mix: 0.1400

The complete data estimate is the average of the P1 and P2 estimates; in the mixed configuration, M1 and M4 have the same segregation type $ab \times cd$, so here the estimate is also the average.

D. The pairwise recombination frequency estimates between markers M1 and M2 in the four datasets are:

Complete data: 0.0550
P1:   0.0200
P2:   0.0900
Mix: 0.0200

The complete data estimate is the average of the P1 and P2 estimates; in the mixed configuration M1 has segregation type $ab \times cd$, whereas M2 has $lm \times ll$, so only the recombination in P1 can be observed.

E. The pairwise recombination frequency estimates between markers M4 and M5 in the four datasets are:

Complete data: 0.0750
P1:   0.0200
P2:   0.1300
Mix: 0.0782

The complete data estimate is the average of the P1 and P2 estimates; in the mixed configuration, M4 has segregation type $ab \times cd$, whereas M5 has $hk \times hk$. The pairwise estimation uses a maximum likelihood model, which results in an estimate slightly different from the average; this is because the actual segregation is slightly different from the Mendelian expectations (1:1:1:1 and 1:2:1).

F. The final results are similar. However, with the regression algorithm, you do not see the differences in recombination between the parents, which you do see with the maximum likelihood method. The markers segregating in only one parent (M2 and M3) do show a higher average χ^2 contribution and nearest-neighbour fit.

Index

accuracy, 36, 136
advanced intercross lines, 38
AIL, 38
allele, 2, 13
allele dosage, 103
allogamy, 101
allopolyploid, 18, 22, 126
allosomes, 19
anchor locus, 117
ant colony optimization, 95
autogamy, 101
autopolyploid, 18
autosomes, 19

backcross, 4, 23
backcross inbred lines, 38
BC_1, 4, 23
BIL, 38
biochemical technique, 4
bridge locus, 117
bridge marker, 107
brute force search, 87

cell division, 10
centiMorgan, 2, 61
centromere, 11
chiasma, 11
chiasmata, 11
chi-square goodness-of-fit, 82, 135
chromatid, 11
chromatid interference, 15, 124
cis-configuration, 22
cM, 2, 61
co-dominance, 22
cold spots of recombination, 8
combinatorial optimization, 90
common logarithm, 49
contingency table, 46
coupling phase, 22, 31, 33, 34, 105
crossover, 11, 14, 124
crossover interference, 17, 62, 68, 72, 124

D, 49
degrees of freedom, 49
deletion, 125
deviance, 27, 49

DH, 4
dioecious species, 101
diploid, 9, 11
disomic, 18
distance geometry, 96
dominance, 4, 22, 23, 33, 36, 103, 127
dominant, 22
dominant allele, 33
dominantly observed loci, 94
dosage, 103
double heterozygote, 31
doubled haploids, 4, 29

egg cells, 10
EM algorithm, 31, 71, 110
epistasis, 22
ES, 94
evolution strategy, 94
evolutionary algorithm, 94
experimental population, 3

F_1, 4, 13, 24
F_2, 4, 29
F_2 backcross, 24
F_2 intercross, 29
F_2-derived doubled haploids, 38
false positive linkage, 52, 54
Fisher information, 28
full-sib family, 102

G^2, 47, 83
GA, 94
gamete, 10
generalized least squares, 67
genes, 2
genetic algorithm, 94
genetic linkage, 5, 45
genetic map, 2
genetic marker, 2
genetic recombination, 3, 9
genotype, 13, 22, 127
geometric map integration, 119
Gibbs sampling, 93, 116
global optimum, 90
graphical genotypes, 133
greedy algorithm, 91
grouping algorithm, 52

Haldane's mapping function, 61, 68
haploid, 9, 11
haplotype, 13
haplotype combination, 22, 110
hermaphrodite species, 101
heterochromosomes, 19
heterogametic, 19
heterozygous, 3, 13
heuristic, 90
hierarchical cluster analysis, 53
homeologous, 18
homogametic, 19
homologous, 11
homologues, 11
homozygous, 3, 13
hot spots of recombination, 8

immortalized population, 4, 29, 38
inbred line, 3
inbreeding depression, 101
incomplete information, 32
independent assortment, 12, 13, 46
independent segregation, 46
insertion, 125
integrated map, 118
interchromosomal recombination, 13
interference, 15, 17, 62, 68, 72, 124
intrachromosomal recombination, 14, 124
inversion, 117
iterative solution, 33

Kosambi's mapping function, 62, 68

L, 26
ℓ, 26
least squares, 66
life cycle, 9
likelihood, 26, 78
likelihood function, 26
likelihood ratio, 27
likelihood ratio chi-square, 47
linear regression, 66, 118
linkage, 17
linkage analysis, 2
linkage group, 5, 45
linkage phase, 22, 102, 127
linkage phase combination, 111
linked, 17
local optimum, 91
loci, 2, 13
locus, 2, 13
LOD, 27
logarithm of odds, 27
log-likelihood, 26, 78, 95, 135
LR, 27

map function, 61
map integration, 117
mapping function, 61
marker, 2
mating types, 10

maximum likelihood, 26, 69, 110, 119
maximum likelihood mapping, 135
meiosis, 9
Mendel's law of independent assortment, 12, 124
Mendel's law of segregation, 13, 124, 130
Mendel's laws, 12
Mendelian inheritance, 124
Mendelian ratio, 130
minimum spanning tree, 92
mitosis, 9, 10
mitotic cell division, 10
ML estimator, 27
model assumptions, 124
model fit, 135
molecular genetic marker, 2
molecular marker technique, 4
MST, 92
multinomial coefficient, 26
multipoint estimate, 71, 89
multipoint estimation, 116
multipoint maximum likelihood, 69, 71, 94, 116

N, 24
natural logarithm, 49
nearest neighbour algorithm, 91
nearest neighbour fit, 136
nearest neighbour stress, 134
non-recombinant haplotype, 14
non-sister chromatid, 11, 124
null-allele, 4, 33, 103

outbreeding species, 6, 23

P_1, 24
P_2, 24
pairwise estimate, 71
pairwise recombination frequency, 110, 131
parental haplotype, 14
parental specific recombination, 113
Pearson chi-square, 47, 130
phase combination, 111
phase type, 104
phenotype, 21, 127
phenotype probability, 133
physical distance, 18
physical linkage, 45
physical map, 7
physically linked, 5
plausible map orders, 136
ploidy level, 9
polyploid, 18
polysomic, 18
post-mapping diagnostics, 123, 132
pre-mapping diagnostics, 123, 129
principal coordinate analysis, 55, 130
pseudo-autosomal, 19
pseudo-testcross, 106
Punnett square, 29
pure line, 3, 23

r, 17
R, 25
rate of recombination, 5, 17, 21
recessive, 22
recessive allele, 33
recombinant haplotype, 14
recombinant inbred lines, 4, 37
recombination fraction, 17
recombination frequency, 17, 21
regression mapping, 135
repulsion phase, 22, 32, 33, 35, 105
RF, 17
RI, 4
RIL, 4, 37

s, 30
S, 32
SA, 92
SALOD, 80
SARF, 81
segregation distortion, 51, 130, 132
segregation ratio, 129, 132
segregation type, 102
self-fertilization, 101
self-incompatibility, 101
self-pollination, 101
self-sterility, 101
sex chromosomes, 19
sexes, 10
sex-linked, 19
simulated annealing, 92
single nucleotide polymorphism, 4
single seed descent, 37

sister chromatid, 11
skewed segregation, 130
smart search algorithms, 88
SNP, 4
spatial sampling, 93
sperm cells, 10
spindle fibres, 11
spurious linkage, 52, 54, 55, 130
SS_R, 82
stress, 134
strongest cross-link, 55
sum of adjacent recombination frequencies, 92, 93, 94, 95
synapsis, 11, 117

tabu search, 93
tetrasomic, 18
three-point testcross, 77
trans-configuration, 22
translocation, 117, 125
travelling salesman problem, 88
two-point estimate, 71
two-point recombination frequency, 131
two-way pseudo-testcross, 106

weighted least squares, 66
weighted linear regression, 91

X^2, 47
X-chromosome, 19

Y-chromosome, 19

zygote, 9

Printed in the United States
By Bookmasters